BTEC National for IT Practitioners:
Systems Units

BTEC National for IT Practitioners: Systems Units

Core and Specialist Units for the Systems Support Pathway

Sharon Yull

LONDON AND NEW YORK

First published 2009 by Elsevier Ltd

Published 2019 by Routledge
52 Vanderbilt Avenue, New York, NY 10017
2 Park Square Milton Park, Abingdon Oxon OX14 4RN

Routledge is an imprint of the Taylor & Francis Group, an informa business

British Library Cataloguing in Publication Data
A catalogue record for this book is available from the British Library

Library of Congress Cataloging-in-Publications Data
A catalog record for this book is available from the Library of Congress

ISBN13 - 9780750686532 (hbk)

Dedication

I would like to dedicate this book to my daughter who is a constant source of inspiration.

Contents

Preface

Introduction

Welcome to the ever changing world of systems and information and communications technology (ICT). This book has been designed to provide you with a range of knowledge, information and skills that will facilitate you in understanding the BTEC Nationals IT Practitioners qualification.

About the BTEC National Certificate and Diploma

The BTEC National Certificate and Diploma qualifications have been designed to provide you with a range of practical skills and underpinning knowledge that will allow you to progress onto a higher level course or prepare you for a job in ICT and computing.

ICT is such a growing area that you will find all sections of the BTEC specification appropriate. The units have been designed with the support of practitioners, experts in the field and also in collaboration with industry. You will be able to use elements of the qualification in a range of situations, whether it is designing a database, managing a network, analysing and maintaining systems or just having an awareness of the impact that communication systems have on industry.

You do not have to have an extensive knowledge of ICT to embark on the BTEC National qualification. Each of the units provides a good coverage of the subject matter. In conjunction, this accompanying book provides additional support in terms of a range of activities and case studies alongside more comprehensive information that follows the guidelines of the specification.

The range of units available on the BTEC Nationals IT Practitioner qualification is quite diverse. The units provide opportunities for you to study at a very specialist level focussing on maintaining computer systems, network management, IT technical support and database design.

On successful completion of this qualification the progression opportunities are quite varied. You could progress onto a Higher National Diploma, Foundation Degree or a Degree programme. Alternatively, you could apply for careers in the areas of systems ICT or computing.

How to use this book

This book provides a support mechanism for the BTEC Nationals IT Practitioner specifications. A range of core and specialist units have been covered within the text, each chapter providing additional materials, activities and case studies.

Each chapter begins with an overview of the content of the related unit and addresses the learning outcomes. Following on from this each of the main headings provides detailed coverage of the learning outcomes.

The activities have been designed to establish your level of learning and provide further opportunities for you to develop your understanding of a specific topic area or concept, and are devised to be used at an 'individual', 'group' or 'practical' level. They are broken down into a range of tasks that require you to undertake research, develop an understanding, provide an opinion, carry out an activity, discuss and present information.

The question sections provide you with an opportunity to re-visit and refresh your understanding of a previous topic.

In some areas of the book certain terminology is used that you may be unfamiliar with. To support your understanding, sections have been included that provide clarification or a definition of the terms referred to.

The effective process of capturing, processing, analysing and storing data is pivotal to the success or failure of many organisations, with patterns and trends being identified and forecasts being made based on current levels of performance.

Data Analysis and Design

Databases are a very important and almost integral part of many business systems. Databases have a range of functions that provide immense storage, processing and analytical capabilities.

This chapter will introduce you to a range of database concepts and techniques. You will become familiar with the mechanics of databases in terms of how they have evolved from paper-based and flat file systems to fully relational systems. A range of design methodologies will be examined, one example being logical data modelling, which can then be applied to your own database designs.

This chapter will provide you with the knowledge and skills required to support you in your own database analysis and design.

The chapter is structured around the following learning outcomes:

- Know modelling methodologies and techniques.
- Understand the tools and documentation required in a logical data modelling methodology.
- Be able to create a logical data model.
- Be able to test a logical data model.

CHAPTER 1

Know modelling methodologies and techniques

When designing a system you would normally follow some sort of framework or method that will provide a structured walk-through for each of the steps and stages involved. When designing a database there are various approaches and techniques that can be applied to ensure that the design meets the needs of the end-user, that it functions, and that it is dynamic and robust.

Database types

Databases have evolved from users and developers being able to understand the semantics of data sets and communicating this understanding clearly and logically. To facilitate this, a specific data model (or models) can be used as a framework for examining and understanding the entities, attributes and relationships between data sets.

Data models can be broken down into three categories:

- **object-based models**: entity relationship, semantic, object-orientated and functional
- **record-based models**: hierarchical, network and relational
- **physical data models**.

Flat file

A flat file is a database system where each database is stored in a single table. Flat files are files that have no records and no structured relationships.

Hierarchical model

The hierarchical data model is so called because of the way in which the data is arranged. The hierarchical model is based on a tree structure with a single table as the root, with the tables forming the branches as shown in Figure 1.1.

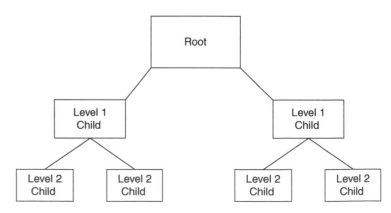

Figure 1.1 Hierarchical data model.

The relationships within this structure are described as parents and children, where a parent can have multiple children but a child can have only one parent. The way in which parents and children are linked together is through the use of 'pointers'. A parent will have a list of pointers extending to each of their children.

The child–parent rule ensures that data is systematically accessible. In terms of navigation, to access a low-level table you would start at the root and work down the tree until you reached the target.

There are a number of problems with the hierarchical structure, including:

- The user must have a good knowledge about how the tree is structured in order to find anything.
- A record cannot be added to a child table until it has already been incorporated into the parent table.
- There will be repetition of data within the database.
- Data redundancy occurs, owing to the fact that a hierarchical database can cope with 1:M relationships but not M:N relationships because a child can have only one parent.

As a result of these problems a different data model was designed to overcome some of the defects attributed to the hierarchical structure.

The network database model

The model originates from the Conference on Data Systems Languages (CODASYL) and was designed to solve some of the more serious problems attributed to the hierarchical database model.

There are similarities between the two models; however, instead of using a single-parent tree hierarchy, the network model uses a set theory to provide a tree-like structure. Child tables can have more than one parent, thus supporting many-to-many relationships.

The design of the network database looks like several trees that share branches, so that children can have multiple parents and vice versa as shown in Figure 1.2.

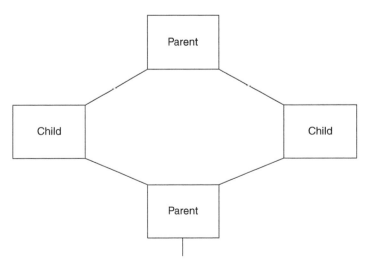

Figure 1.2 Network model structure.

Although an improvement on the hierarchical model the network model still had some intrinsic problems. The major problem was that the model was difficult to implement and maintain, most implementations being used by computer programmers and not end-users. A less complex database model was required that could be used by real end-users, thus giving birth to the relational database model.

The relational database model

The relational database model was developed from the work carried out by Dr E. F. Codd at IBM in the late 1960s. This new model looked for ways to solve the problems with the existing models.

Central to the relational database model is the concept of a table (relation) in which data is stored. Each table is made up of records (tuples) and fields (attributes). Each table has a unique name that can be used by the database to find the underlying table. Unlike previous models that have a defined hierarchy there is no specific navigational mapping. The relational model works on the basis that any manipulation of data is carried out via the data values themselves. Therefore to retrieve a row from a table you would compare the value stored within a particular column for that row with some search criteria.

An example of this can be seen in the following. If you wanted to search for all the rows from the 'Flight' table that had 'Paris' in the 'destination' column the database might return a list as shown in Table 1.1.

Table 1.1 Flight table query

Destination	Flight code	Date	Airline
Paris	P-45778-9	17 July 07	TEY Flights
Jersey	J-28399-9	17 July 07	TEY Flights
Manchester	M-45273-5	17 July 07	TEY Flights
Edinburgh	E-38400-2	17 July 07	TEY Flights

Data from a retrieved row can then be used to query another table. For example, you may want to know whether this flight arrived on time or if it was delayed.

By using the 'flight code' from this query as the keyword in the new query, you could look in the 'arrivals table' and look for the row where the flight number is 'P-45778-9' to check the arrival time.

This query methodology makes the relational model a lot simpler to understand. Other benefits of this model include:

- It provides useful tools for database administration (the tables can store actual data and be used as a means for generating meta-data).
- The use of tables allows for structural and data independence.
- It allows for more effective design strategies.
- It has a logical rather than a physical representation.

All of these benefits have secured the relational database as being a firm favourite within academic, medical, government, financial, commercial and industrial domains.

Data model – defines data objects and the relationships between them.
Entity – within a database environment an entity is an object such as a table or a form.
Tuple – can mean different things depending on the environment in which you are working. For example, in a programming language environment, a tuple is an ordered set of values. In a database environment, a tuple is similar to a record in non-relational databases.

Activity 1.1

1. Provide a short summary outlining these three data model categories and the three elements that they comprise.

2. Compare and contrast the following data models:

    ● hierarchical
    ● network
    ● relational.

3. Using diagrams, provide a visual representation of a hierarchical, a network and a relational data model.

Modelling methodologies

A range of modelling techniques can be used in database design; some of these can be classed as top–down and others as bottom–up.

Logical data modelling

Object-based models focus on concepts such as entities, attributes and relationships. The logical data model encompasses these concepts and thus emerges as one of the main techniques in the design of conceptual databases. The object-orientated model builds upon the foundations of the logical data model by describing actions associated with an object and how it behaves.

The diagrammatic aspect of logical data modelling (LDM) is referred to as the entity relationship diagram (ERD), as shown in Figure 1.3. The ERD has four main components:

● entities
● relationships
● degree
● optionality.

Entities

Entities provide the source, recipient and storage mechanism for information that is held on the system. They are distinct objects (people, events, places, things or events) that are represented within a database. For example, typical entities for the following systems include:

Library system
Entities:

● Book
● Lender
● Reservation
● Issue
● Edition

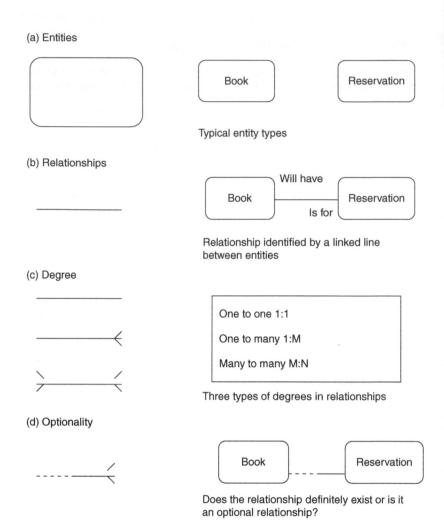

Figure 1.3 Entity relationship diagram tools.

Hotel system
Entities:

- Booking
- Guest
- Room
- Tab
- Enquiry

Airline system
Entities:

- Flight
- Ticket
- Seat
- Booking
- Destination

Each entity will have a set of attributes, the attributes describing some aspect of the object that is to be recorded. For example, within the library system, an object exists – book; and the ways used to describe the book/object are the attributes.

Entity:	Book
Attributes:	ISBN number
	Title
	Author
	Publisher
	Publication date

Each set of attributes within that entity should have a unique field that provides easy identification to the entity type. In the case of the entity type 'book' the unique key field is that of 'ISBN number'. The unique field or 'primary key' will ensure that although two books may have the same title or author, no two books will have the same ISBN.

Relationships

A relationship provides the link between entities. The relationship between two entities could be misinterpreted; therefore labels are attached at the beginning and at the end of the relationship link to inform parties exactly what the nature of the relationship is. For example, if you had two entities, book and author, linked as illustrated in Figure 1.4, the nature of the relationship could be any of the following:

Figure 1.4 Example of entity relationships.

- An author can write a book, therefore the book belongs to an author.
- An author can refer to a book, therefore the book is in reference by an author.
- An author can buy a book, therefore the book is bought by an author.
- An author can review a book, therefore the book is reviewed by an author.

The actual relationship that exists in this scenario is that an author reviews a book and the book has been reviewed by an author, as shown in Figure 1.5.

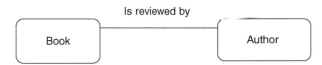

Figure 1.5 Relationship descriptor.

Degree

There are three possible degrees to any entity relationship, as shown in Figure 1.6.

- **One to one (1:1)** – only one occurrence of each entity is used by the adjoining entity.
- **One to many (1:M)** – a single occurrence of one entity is linked to more than one occurrence of the adjoining entity.
- **Many to many (M:N)** – many occurrences of one entity are linked to more than one occurrence of the adjoining entity.

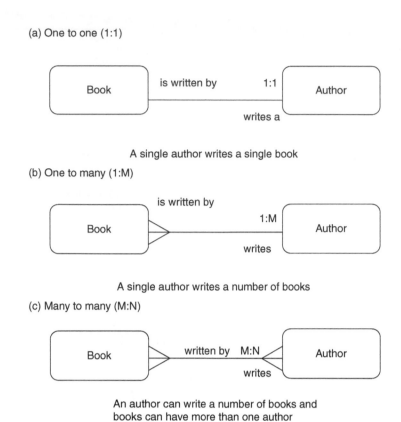

Figure 1.6 Degrees of relationships: (a) one to one; (b) one to many; (c) many to many.

Although M:N relationships are common, the notation of linking two entities directly is adjusted and a link entity is used to connect the two:

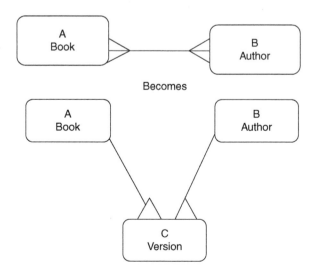

In this scenario an author can write a number of versions of a book, and a book can have a number of versions that have been written by an author.

Optionality

There are two status types given to a relationship; first those that definitely happen or exist, and secondly those that may happen or exist, this second status being referred to as 'optional'.

A dashed rather than a solid link denotes optionality in a relationship. In this scenario an author may or may not decide to write a book:

LDM provides a detailed graphical representation of the information used within a system and identifies the relationships that exist between data items.

Activity 1.2

1. For each of the following systems identify at least six appropriate entities:

   - college
   - cinema
   - bank
   - flight booking.

2. Draw an entity relationship diagram to represent each of these systems. What problems, if any, have occurred?

3. For each of the following systems complete either the entity name, relationship type or both:

   - holiday system

   - video/DVD rental system

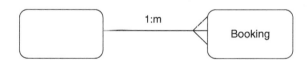

   - restaurant system

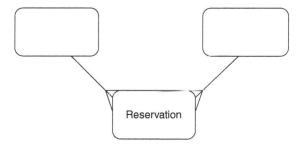

● **library system.**

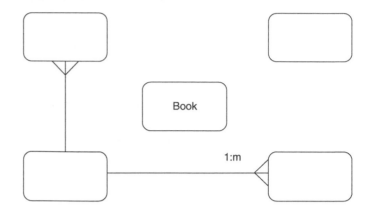

4. **For the following entities, provide a complete attribute list with a suitable primary key:**

● **customer**
● **student**
● **patient**
● **resort**
● **transaction**
● **reservation.**

5. **Why is it important to have a primary key for each entity?**

Normalization

Normalization is a bottom–up approach to database design that starts with the examination of relationships between attributes. E. F. Codd (1972) first developed the process of normalization. The initial framework was modelled on three stages or tests that were applied to a given 'relation'. These stages extended from:

● first normal form (1NF)
● second normal form (2NF)
● third normal form (3NF).

An improved version of 3NF was later developed by R. Boyce and E. F. Codd in 1974 and is referred to as Boyce–Codd Normal Form (BCNF). In 1977 and 1979 higher normal forms, fourth (4NF) and fifth (5NF), were introduced by Fagin; however, these deal with practical situations that are quite rare.

An example of normalizing data up to third normal form is presented in Tables 1.2–1.5.

The stage of converting an unnormalized data set into first normal form (1NF) involves looking at the structure as all records must be of fixed length. To address this, the data needs to be divided into two (Table 1.2):

Table 1.2 Unnormalized data set

Student number[a]	Student name	Module code	Module name	Grade	Lecturer	Room number
CP123/OP	Greene	C122	IS	D	Jenkins	B33
CP938/CP	Jacobs	C123	Hardware	M	Smith	B33
CF489/LP	Browne	C124	Software	M	Osborne	B32
CP311/CP	Peters	C111	Internet	P	Chives	B32
CR399/CP	Porter	C110	Web design	D	Crouch	B33
CD478/JP	Graham	C107	Multimedia	M	Waters	B31
CR678/LP	Denver	C106	Networking	P	Rowan	B31

[a]Primary key.

- **f** xed part
- variable part that contains repetitions.

To rejoin the data the variable part must contain a key from the fixed part to make a composite key of Student number/Module code as shown in Table 1.3.

Table 1.3 First normal form (1NF)

Student number	Student name
CP123/OP	Greene
CP938/CP	Jacobs
CF489/LP	Browne
CP311/CP	Peters
CR399/CP	Porter
CD478/JP	Graham
CR678/LP	Denver

Student number	Module code	Module name	Grade	Lecturer	Room number
CP123/OP	C122	IS	D	Jenkins	B33
CP938/CP	C123	Hardware	M	Smith	B33
CF489/LP	C124	Software	M	Osborne	B32
CP311/CP	C111	Internet	P	Chives	B32
CR399/CP	C110	Web design	D	Crouch	B33
CD478/JP	C107	Multimedia	M	Waters	B31
CR678/LP	C106	Networking	P	Rowan	B31

Problems still exist with 1NF that call for examining any partial dependencies that exist within the data set. Any non-key field that is dependent on (can be derived from) only part of the key is said to be partially dependent.

For example, Module name is derived from the Module code (it makes no difference which student is taking that module).

However, the grade is not partially dependent because it is the grade for an individual student taking a particular module.

Problems with partial dependence:

- **Updating** – if for example the name of module C124 (Software) changed to (Operating systems) every record that contained details of this old module would need to be updated
- **Insert anomaly** – if a new module needs to be added under the current organization the module details, e.g. name, are stored with records about individual students taking that module. A new module would have no students enrolled onto it, so where would it be stored?
- **Deletion anomaly** – if a student decides to drop a module or leave the course all the details of that module will be lost.

The next step for third normal form (3NF) conversion (Table 1.4) is to ensure that the data does not show any indirect dependencies. All non-key fields should be defined by the key directly and not by another non-key field.

Table 1.4 Second normal form (2NF)

Student number	Student name
CP123/OP	Greene
CP938/CP	Jacobs
CF489/LP	Browne
CP311/CP	Peters
CR399/CP	Porter
CD478/JP	Graham
CR678/LP	Denver

Student number	Module code	Grade
CP123/OP	C122	D
CP938/CP	C123	M
CF489/LP	C124	M
CP311/CP	C111	P
CR399/CP	C110	D
CD478/JP	C107	M
CR678/LP	C106	P

Module code	Module name	Lecturer	Room number
C122	IS	Jenkins	B33
C123	Hardware	Smith	B33
C124	Software	Osborne	B32
C111	Internet	Chives	B32
C110	Web design	Crouch	B33
C107	Multimedia	Waters	B31
C106	Networking	Rowan	B31

For example, the room number is defined by the lecturer and not by the module. This creates similar dependency problems of updating and insertion and deletion anomalies.

- If a lecturer moves to another room their room location would also need to be changed.
- A lecturer not allocated to teaching a module cannot have any room details stored.
- If a lecturer stops teaching a module the room details will be lost.

To overcome these problems further divisions in the data structure are required (Table 1.5).

Table 1.5 Third normal form (3NF)

Student number	Student name	
CP123/OP	Greene	
CP938/CP	Jacobs	
CF489/LP	Browne	
CP311/CP	Peters	
CR399/CP	Porter	
CD478/JP	Graham	
CR678/LP	Denver	
Student number	**Module code**	**Grade**
CP123/OP	C122	D
CP938/CP	C123	M
CF489/LP	C124	M
CP311/CP	C111	P
CR399/CP	C110	D
CD478/JP	C107	M
CR678/LP	C106	P
Module code	**Module name**	**Lecturer**
C122	IS	Jenkins
C123	Hardware	Smith
C124	Software	Osborne
C111	Internet	Chives
C110	Web design	Crouch
C107	Multimedia	Waters
C106	Networking	Rowan
Lecturer	**Room number**	
Jenkins	B33	
Smith	B33	
Osborne	B32	
Chives	B32	
Crouch	B33	
Waters	B31	
Rowan	B31	

Object-orientated analysis and design

Object-orientated analysis and design (OOAD) examines systems from an 'object' perspective and how these objects interact with each other. Objects are characterized by their class and state or data elements and their behaviour. OOAD focuses on what the system does and how it functions, and also examines how the system functions.

Understand the tools and documentation required in a logical data modelling methodology

LDM is one methodology that can be used in database analysis and design. LDM consists of a set of tools that are used to create the systems model and the accompanying documentation to support the design and justify the techniques used.

Logical data modelling concepts and constraints

There are two components to LDM. The set of tools used to create an ERD (as referred to in Modelling methodologies, above) form the first component. The joining together of entities with a link relationship forms the ERD (Figure 1.7).

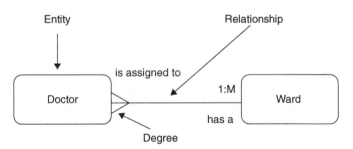

Figure 1.7 Entity relationship diagram.

A doctor is assigned to a ward, and a ward can have a number of doctors assigned to it.

Doctor	Ward
Doctor ID*	Ward number*
Name	Name
Department	Description
Bleep number	

*Denotes a primary key.

The second component of LDM is the documentation that supports the ERD. The documentation that is used includes:

- entity descriptions
- attribute lists.

Every entity should have an associated 'entity description' which details items such as:

- entity name and description
- attributes
- relationship types and links.

Every entity has a set of attributes. If a large system is being investigated a number of entities and their associated attributes will need to be defined, therefore an 'attribute list' can be prepared. Attribute lists identify all of the attributes and a description of the attributes. The primary key attribute, which is normally made up of numerical data, e.g. supplier number, national insurance number, examination number, is referred to first, followed by the remainder of the attribute items.

Producing entity relationship diagrams

Case Study 1.1

Wrights Supermarkets

Wrights Supermarkets is an established chain of supermarkets that are located across the country. Over the past six months the managing director of the chain, Gabriel Johnson, has discovered that they are losing their proportion of the market to another competitor. Since the beginning of the year their market share has fallen from 16 per cent to 12 per cent.

Wrights has fifteen stores across the region, all located in major towns or cities. The structure of the company is very hierarchical, with the lines of command being generic across all branches as shown in Figure 1.8.

All of the branches communicate on a regular basis. Branches also distribute surplus stock items to other branches if they are running low, to reduce supplier ordering costs.

All of the functional departments are located at the head office, which has the following implications for each branch:

- All recruitment is done through head office for each of the branches, which means that all the application forms have to be sent either by post or online (if the application was filled in online).
- All stock ordering is done through head office, who have negotiated local supplier contracts for each of the branches.
- All of the promotions, for example 'buy one get one free', and all of the price reductions and special offers are filtered through from sales at head office.
- All salaries are paid via the finance department at head office.
- All deliveries and distribution are made through local suppliers in conjunction with head office instructions.

East Anglia branch

Mr Johnson has asked for an investigation to take place based at the branch in East Anglia to identify what problems exist, with proposals on how they can be addressed. The focus of the investigation is the 'fresh produce' department.

Fact-finding

Using a variety of fact-finding techniques the following information has been gathered:

1. There are 150 employees at the branch:

   - Store Manager – Mr Howell
   - Deputy Manager – Ms Peters
   - five store managers, five assistant managers and five supervisors
   - fifty full-time and part-time checkout staff
   - forty full-time and part-time shelf-stackers
   - ten stock clerks
   - ten trolley personnel
   - three car park attendants
   - twenty cleaners, gardeners, drivers and other store staff.

2. Each of the store managers control their own areas, with their own shelf-stackers and stock personnel.
3. All stock ordering is batch processed overnight to head office on a daily basis by each of the store managers in consultation with the deputy branch manager.
4. All fresh produce is delivered on a daily basis. Non-perishable goods are delivered three times a week by local suppliers.
5. All bakery items are baked on site each morning.

Fresh produce system

Jonathon and his assistant manager, Margaret, manage the fresh produce department. Within the department their supervisor, John, oversees six display/shelf-stackers and four stock personnel.

After consultation with a range of employees the following account of day-to-day activities has been given.

Each day Margaret holds a staff meeting within the department to provide information about new promotions, special discounts or stock display arrangements.

Any information regarding new promotions comes through from head office. All information received regarding promotions, etc., is filed in the branch promotions file. If any price adjustments need to be made that day the stock personnel are informed to check the daily stock sheets.

After the meeting the stock personnel liaise with the shelf/display personnel with regard to new stock that needs to go out onto the shopfloor. The information about new stock items and changes to stock items comes from the daily stock sheet. When new items have been put out or stock price adjustments made they are crossed off the daily stock sheet.

Items that have arrived in that day are delivered from the local fresh produce supplier. When the items come in the stock personnel checks the daily stock sheet for quantities and authorizes the delivery. If items have not arrived or there is an error in the order a stock adjustment sheet is filled in, which is kept in the stock office. At the end of the day John will inform Jonathon of the stock adjustments. Jonathon then sends off a top copy of the adjustment sheet to head office and files a copy in the stock cabinet.

Information about stock items running low comes from the daily stock sheet. If an item is low a stock order form is completed. A top copy is sent to head office and a copy is filed in the stock cabinet. Orders should be made five days before the actual requirement of the stock, as head office then processes the information and contacts the local supplier. In an emergency local supplier information is held by Margaret, who can ring direct to get items delivered. This, however, costs the company more money because a bulk order has not been placed. Authorization also has to be given by the operations manager at head office. Information also has to be filled in on the computerized stock request form; this is e-mailed to head office each day, and they then send back confirmation.

Head office dictates that all documents filled in online also need to have a manual counterpart, one which is sent off and the other which is filed with the branch.

Problems with the system

1. Sometimes the network at head office is down, which means that stock items are not received within five days.
2. The promotions are not always appropriate, either because of a lack of certain stock or because the stock of which the branch has a surplus is wasted because they cannot set their own promotions in store.
3. The stock cabinet is filled to capacity and because everything is in date order it is difficult to collate information about certain stock items.
4. If there is an error in the stock delivery nothing can be supplied until the paperwork has been sent off to head office or authorization has been given, even if the supplier has the stock requirement on his lorry.
5. There is too much paperwork.
6. There is little communication with other departments.
7. Targets which are set by head office cannot always be met due to the stock ordering problem.
8. Some stock items that come in are not all barcoded.

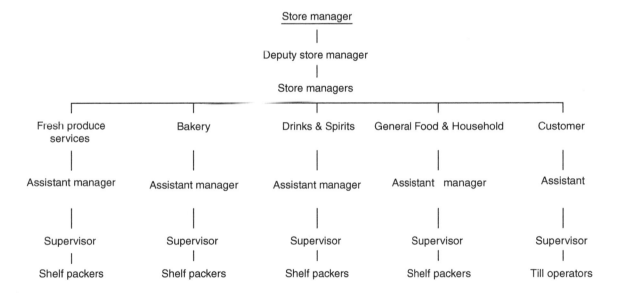

Head office structure

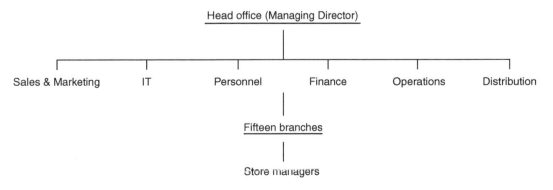

Figure 1.8 Branch structure.

Hungarian notation – a notation widely used in programming, especially 'C', that represents the data type or intended use of a variable within its name. Hungarian notation generates a way to provide standard names for the variables, data structures, procedures and methods created within a software project.

Activity 1.3

1. Using the information from the fresh produce department at Wrights supermarkets produce an entity relationship diagram clearly labelling all relationships.
2. Produce a full attribute list for five of the entities.
3. What enhancements would you make to your ERD after examining some of the problems listed?

There are constraints associated with the use of any methodology and this is especially true of LDM. Constraints associated with this modelling technique include:

- domain
- entity
- referential
- user
- operational (platforms or organizational conventions, etc.).

Activity 1.4

1. Complete the table to provide an example of how each of the constraints can impact on logical data modelling techniques:

Constraint	Example
Domain	
Entity	
Referential	
User	
Operational	

Technical documentation and the purpose of documentation

Documentation is a crucial part of LDM. Documentation is used to support the modelling tools used and to clarify and justify what data exists and how it interacts with other entities within a given system. There is a range of documentation that can be used in LDM, including data dictionaries, and entity descriptions and the notation used can vary depending on the technical nature of the document, for example Hungarian to Polish or 'prefix' notation.

An example of Hungarian notation is:

Prefix	Type	Example
b	boolean	bool bStillGoing;
c	character	char cLetterGrade;
si	short integer	short siChairs;
i	integer	int iCars;
li	long integer	long liStars;
f	floating point	float fPercent;

Polish/prefix notation – a notation used for logic, algebra and mathematics, which is widely used in computer programming. An example of Polish notation is:
Instead of writing 3 + 7 it would become +3 7.

Documentation is important because it can provide an audit trail of any changes made and it can support the user in any maintenance required on the system.

The use of documentation can support users in their work environment, provide clear instructional guidelines on 'how to do' type tasks, and eliminate problems and ease maintenance as self-diagnosis and solutions may be possible from within the documentation. In addition, documentation can ease any alterations or changes that have to be made to a system because there may be a clear audit trail outlining what has been done previously or a test plan and log of problem areas that require resolution.

Be able to create a logical data model

The principal tools and techniques of LDM and ERDs have been analysed in previous sections of this chapter. It is important to understand the tools and techniques that are used in the methodology to create a visual representation of what is happening within a given system.

When a logical data model is being created it is important to recognize and embrace the requirements of the user in terms of data storage, retrieval and validation. The need for clear and supportive documentation is also essential to ensure that accuracy and consistency have been applied throughout.

The use of modelling tools can indeed provide a clear overall view for a given user as to what is happening within a given system. For example, an ERD may identify that there are inadequacies in terms of data storage. It may become apparent that once information is captured within a given system it is not stored appropriately and therefore communication channels or inefficiencies within that system become apparent, as in the example of a college system (Figure 1.9).

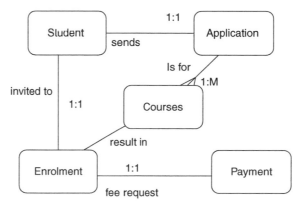

Figure 1.9 College entity relationship diagram.

In this system it is evident that there are problems with the communication and storage of data; for example, the student never receives details about any fee payments. In addition, it is not evident whether there is an approvals process whereby, when the application is received, any checks are put in place to ensure that the student meets the requirements of a particular course. It is also unclear where the final application and enrolment details for this student are stored. The inclusion of additional data such as 'lecturer', 'approval' and 'department' may help to address these issues.

In terms of data retrieval and validation, it is also unclear in this example who accesses the information and what checks have been put in place to make sure that a student meets certain entry requirements or qualifies for a fee waiver.

LDM, once complete and accurate, will provide the framework for further system developments and provide the structure in terms of the required tables and fields for a database solution from which a user can store, retrieve and conduct any analysis and queries required.

Databases carry out a range of functions to support all types of users. Their primary function is to store volumes of data and specific formats to allow for easy processing and access. Data that is input into a database can then be formatted into meaningful and useful information.

In terms of creating a database and inputting data based on the ERD, two types of input forms that can be used are tables and forms.

Tables are set out in a grid/tabular format and provide an overview of the information for a database (single table at a time). The significance of tables is that you can create as many or as few as you require, each table having its own unique data set.

Using the example of the Curious Book Shop, it is easy to see how entities from the ERD convert into tables for a database (Figure 1.10).

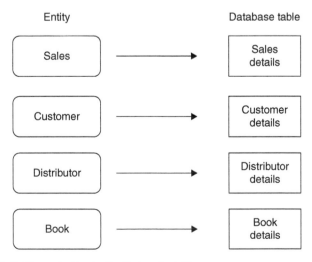

Figure 1.10 Entities and tables for The Curious Book Shop.

Each of the tables can contain details about the information category; therefore the 'book table' might contain information such as:

ISBN number $_{\text{key field}}$
Title
Author

Data types – when entering field names into a table design template you are required to select an appropriate data type, e.g. text, numeric, date/time.

Date of publication
Edition
Publisher
Category

Tables can be created by typing information directly into a tabular template (Figure 1.11), or by setting up the table design by selecting field names and data types (Figure 1.12). Once the structure of the table has been set up, data can be input (Figure 1.13).

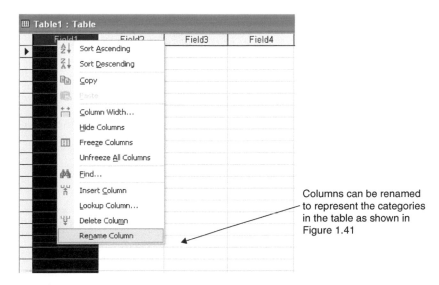

Columns can be renamed to represent the categories in the table as shown in Figure 1.41

Figure 1.11 Creating a table template.

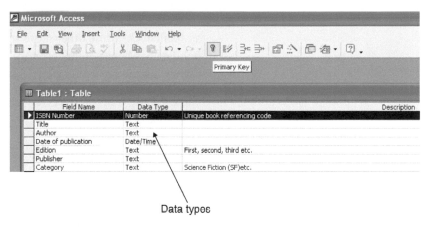

Data types

Figure 1.12 Using a design template.

Field Name	Data Type	Description
ISBN Number	Text	Unique book referencing code
Title	Text	
Author(s)	Text	
Date of publication	Text	
Edition	Text	First, second, third etc.
Publisher	Text	
Category	Text	Science fiction (SF)

All data types have been changed to text in this example; the result of this is shown below.

Figure 1.13 Inputting data.

Identifying, selecting and modifying databases

Databases can be set up specifically to meet an end-user's requirements. At the initial data input stage, data can be entered using a specific data type as previously explored. In addition, field properties such as the length of a field can be changed, the default value and the field name can be renamed in conjunction with other modifications, and validation rules can be set up (Figure 1.14).

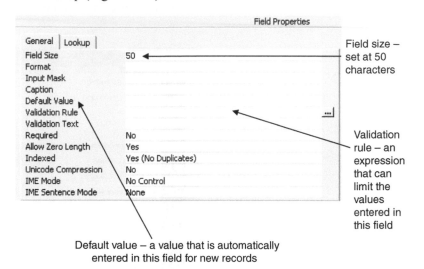

Figure 1.14 Setting up rules and validation procedures.

Checks and validation

A number of checks can be carried out to ensure that data has been input correctly. Some of these checks include the use of general tools such as spellcheck, while other more specific checks can identify whether or not data has been set up correctly, that it is in the right format and that it carries out the required task or function requested.

Table 1.6 provides an overview of some of the more specific checks that can be carried out.

Forms, reports and queries

Forms can be used to provide a more user-friendly method of inputting data. Forms can also be customized to include a range of graphics and background images.

There are two options to designing a form. The first is through the design palette and toolbox (Figure 1.15). The second way is to use the wizard function, which allows you in a few easy steps to select the fields, choose a layout, style and label (Figure 1.16).

Reports provide an output tool that consolidates a set of information based on specific criteria that have been selected by the user, an example of which can be seen in Figure 1.17.

Databases can provide a very dynamic environment for storing, retrieving, querying and outputting information. Databases can be

Table 1.6 Data checks

Check	Purpose
Presence check	To ensure that certain fields of information have been entered, e.g. hospital number for a patient who is being admitted for surgery
Field/format/picture check	To ensure that the information that has been input is in the correct format and combination (if applicable), e.g. the surgery procedure has an assigned code made up of two letters and six numeric digits: DH245639
Range check	To ensure that any values entered fall within the boundaries of a certain range, e.g. the surgery code is only valid for a four-week period (1–4); therefore any number entered over 4 in this field would be rejected
Look-up check	To ensure that data entered is of an acceptable value, e.g. types of surgery can only be accepted from the list orthopaedic, ENT (ears, nose and throat) or minor
Cross-field check	To ensure that information stored in two fields matches up, e.g. if the surgeon's initials are DH on the surgery code they cannot represent the surgeon Michael Timbers, only Donald Hill
Check digit check	To ensure that any code number entered is valid by adding in an additional digit that has some relationship with the original code
Batch header checks	To ensure that records in a batch, e.g. number of surgeries carried out over a set period, match the number stored in the batch header

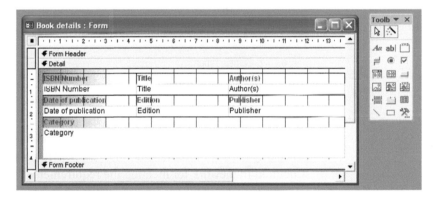

Figure 1.15 Form design view.

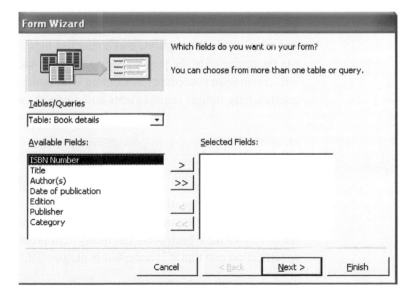

Figure 1.16 Form wizard.

Book details

ISBN Number	1-2300-45667
Title	Happy Holidays
Author(s)	Julia Lodge
ISBN Number	1-1100-45569
Title	Beyond space
Author(s)	Arthur Moon
ISBN Number	1-5678-00011
Title	Coming Home
Author(s)	Margaret Housey
ISBN Number	1-3423-39940
Title	The Ultimate Guide to Shoes
Author(s)	Jonathon Heel
ISBN Number	1-2738-92738
Title	History Through The Ages
Author(s)	Paul Battle

Figure 1.17 Sample report.

produced using a top–down approach or a bottom–up approach to design; either way the use of normalization or LDM will provide an efficient and structured framework for meeting the requirements of users.

Documentation

A range of documentation can be used to support LDM and design. Some documentation supports the actual ERD structure, such as entity and relationship descriptions and constraint lists. Other documentation can be attributed to the physical design stages, such as setup and implementation procedures associated with tables, relationships and queries, data storage requirements and data validation procedures.

Be able to test a logical data model

Testing can be carried out at any stage as an incremental process or at the end of implementation. Whenever testing is carried out the function remains the same; the need to instil quality and checks that the user requirements have been met, that the system proposal meets the required specification and that it is accepted is paramount.

Testing can fall under the category of integrity tests, which look at the entity, relationship, field and constraints, or error testing, which identifies whether a data set falls within the categories of normal, erroneous or extreme.

Testing will ensure that components operate effectively, uniformly and consistently at all levels and a test plan can provide the documentary evidence to support this. Points to consider when testing include:

- version control – what test number/version you are up to
- what has been tested
- at what stage in the development
- the purpose of the test
- the results of the test
- comments – did the test run as expected?

Each time a test is carried out the test plan should be updated and used as a working document that can be integrated into the final evaluation. A sample test plan can be seen in Table 1.7.

Table 1.7 Sample test plan

Test ID	Description	Expected result	Actual result	Pass/ fail
1	Application start/exit Application starts correctly	Display application menu screen	Menu screen loads	P
2	All menu screen options/ buttons function correctly	Options cause correct screens to be displayed	All options OK	P
3	Application closes down correctly	Application closes	Application fails to close	F

Questions and review

1. Identify a range of database types.
2. Describe the main features of a hierarchical database.
3. A number of modelling methodologies can be used in the database design process. Identify two different methodologies and describe what they do and/or what they show.
4. What are the key concepts associated with logical data modelling?
5. What constraints are associated with logical data modelling?
6. What types of technical documentation might be required when undertaking logical data modelling?
7. What is the purpose of having documentation?
8. What factors should be taken into consideration when creating a logical data model?
9. Testing can take place in various ways. Integrity and error testing are examples of this. What features would be tested if undertaking integrity and error tests?
10. What criteria or features should you consider when devising a test strategy?
11. What, in your opinion, is the purpose of having test documentation?

CHAPTER 1

Assessment activities

Grading criteria	Content	Suggested activity
Pass		
P1	Describe the advantages and disadvantages of the specified database types, using examples.	Produce a table of examples that describes the advantages and disadvantages of specified database types.
P2	Describe the advantages and disadvantages of two analysis and design methodologies, using examples.	Expand the table to include the advantages and disadvantages of two analysis and design methodologies such as LDM, normalisation or OOAD etc.
P3	Identify and describe potential modelling constraints that could arise from a logical data model, using examples.	Produce a short report that identifies and describes potential modelling constraints that could arise from a logical data model and the concepts of LDM. You should also explain the benefits and constraints of the logical modelling process.
P4	Describe the concepts involved in logical data modelling.	Add a section to the report that includes a description of the concepts involved in LDM.
P5	Produce a data model to meet specified user requirements.	Produce a data model to meet specified user requirements. Once the data model has been created the data modelling process can then be evaluated for effectiveness in terms of meeting the user requirements.
P6	Produce a test strategy and test plan for normal situations for a data model.	Produce a test plan and a test strategy that outlines 'normal situations' for the data model produced in P5.
P7	Implement a logical data model.	In conjunction with P5, a logical data model should be implemented.
Merit		
M1	Explain the benefits of the logical data modelling process.	In conjunction with P3, a section within the report could explain the benefits of the LDM process.
M2	Explain the constraints developed in a logical data model to meet specified user requirements.	In conjunction with P3, the final part of the report could identify and explain the constraints developed in a LDM. You could also draw upon your practical experiences from P5 and P7.
M3	Justify the requirements for all types of test required to ensure a logical data model is efficient and effective.	In conjunction with M4, justify the requirements for all types of test required to ensure a logical data model is efficient and effective. You should also provide a written justification as to the purpose of complete and accurate technical documentation for a logical data model and associated testing
M4	Justify the purpose of complete and accurate technical documentation for a logical data model and associated testing.	In conjunction with M3, justify the requirements for all types of test in order to ensure that a logical data model is efficient and effective. You should also provide a written justification as to the purpose of complete and accurate technical documentation for a logical data model and associated testing.
Distinction		
D1	Evaluate the effectiveness of the data modelling process in producing an efficient data model to meet user requirements.	In conjunction with P5, once a data model has been designed, a short written explanation can be provided that evaluates the effectiveness of the process in producing an efficient data model to meet user requirements.
D2	Evaluate a model produced against an initial brief, and suggest improvements to enhance the model to meet user requirements.	Evaluate the model produced for P5 and suggest improvements to enhance it to meet user requirements.

Courtesy of istockphoto, Dvorjakusan, Image#7992379

Human computer interaction (HCI) and the technologies associated with it affect everybody's life, even to the extent that a good user interface can actually drive technological change.

Human Computer Interaction

Human computer interaction (HCI) examines and analyses the way in which users interact with computers. HCI is quite a broad category that covers a number of disciplines. In terms of programming and design HCI is paramount to ensure that systems are developed in the right way and that they are fit for the purpose or user for which they were intended. From an end-user support view, HCI can ensure that the learning experience is more enjoyable and that the adaptability period is reduced because of familiar screens, menus or other interface tools.

This chapter will examine a range of HCI theory that will enable you to understand why systems are designed in a certain way, the fundamental process of interface design, and the effectiveness of various input and output mechanisms. In addition the social aspects of HCI will be introduced to enable you to appreciate how this can impact on society, the economy and culture.

The chapter is structured around the following learning outcomes:

- Know about the impact of HCI on society, the economy and culture.
- Understand the fundamental principles of interface design.
- Be able to design and produce simple interactive input and output based on HCI principles.
- Be able to compare and contrast, using HCI principles, the effectiveness of different designs of input and output.

CHAPTER 2

HCI has had a tremendous impact on society, the economy and culture. It has influenced the way in which systems are designed and how information and data are displayed and accessed. HCI has provided developers, programmers and designers with 'food for thought', enabling them to think about what the function of a particular tool, system or application is and how this can be conveyed to a user in a way that is understood, recognized and visually appealing.

Development

HCI involves the study of methods for designing the input and output screens for a given system to ensure that it is user friendly. A number of factors should be taken into consideration in terms of HCI development, as identified in Figure 2.1. The way in which users interact with systems is dependent on a range of factors, as shown in the figure:

- organizational
- health and safety
- task
- system functionality
- productivity.

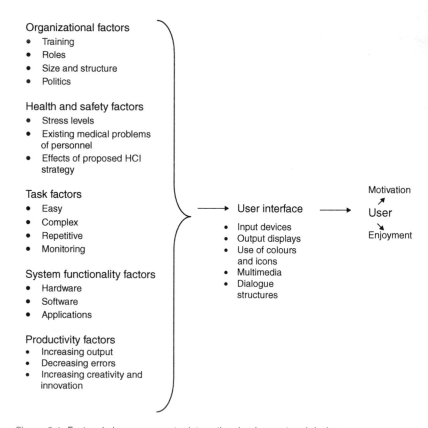

Figure 2.1 Factors in human computer interaction development and design.

However, other factors that are not so defined should also be taken into consideration. These factors in some cases are more important in HCI because they examine the more humanistic elements such as cognition and perception, motor performance, personality and culture.

Therefore, developments and design should take into account end-users':

- memory (short, medium and long term) and learning
- problem-solving and decision-making abilities
- attention span, perception and recognition
- anxieties and fears
- age and gender
- response and stimuli
- vision
- hearing
- touch and coordination
- physical strength
- personality
- disabilities
- awareness of culture and international diversity (customs, etiquette, formalities and tradition).

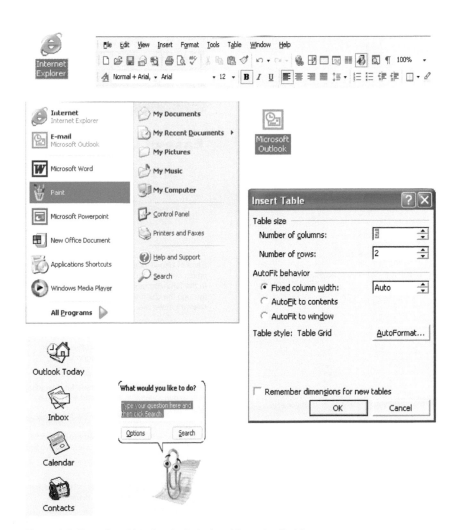

Figure 2.2 Examples of interface tools designed for novice/first time users.

Certain measures can be taken to address each of these. In terms of the psychological factors a system can be designed that:

- is user friendly
- provides support and help to novice users
- provides shortcuts for more advanced expert users
- makes use of human long-term memory to maximize efficiency.

Users can be broken down into four categories. The profile of each user type has an impact on HCI and the type of interface that needs to be developed, for example a user who is considered to be a 'novice' will not be familiar with the layout or functions of a particular interface, therefore any interface needs to be simple, clear, instructive and possibly graphical. Examples can be seen in Figure 2.2.

Some organizations such as IBM and Microsoft provide specific solutions for software and its ease of use, realizing that systems should be 'user driven' in conjunction with being 'functionally driven' (Case Study 2.1).

Case Study 2.1

IBM – solutions for usability

Ease of Use

Attribute Explorer	NotesBuddy	Systems Journal
rapid data analysis through attribute bar chats	e-mail and instant messenger	featuring ease of use topics

Ease of Use is vital to the success of most products and services. The user experience directly affects sales, service cost, productive use, customer loyalty and almost every other aspect of doing business. In the following seven sections, this site addresses the challenge of creating great user experiences through the discipline of User Engineering, supported by design guidelines, tools, services and other relevant materials.

Value
Discover a compelling value proposition for improving your total user experience.

User Engineering
Learn about the definitive process for designing user experiences that satisfy and exceed user expectations.

Services
Find out how IBM's experienced professionals can assist you in designing outstanding products and solutions.

Downloads
Try out the various applications, resources and UCD tools that will help improve usability.

Journal
Get the latest information on ease of use from IBM and featured companies. Subscribe to the monthly newsletter.

Design
Explore design principles and guidelines for Web sites, desktops, 'out-of-box' and other common experiences.

Conference
The annual Make IT Easy conference is a forum for the exchange of ideas and information on ease of use with IT professionals from around the world.

http://www-306.ibm.com/ibm/easy/eou_ext.nsf/publish/558

CHAPTER 2

User-centred design – design built around a user's needs and their working environment.
System-centred design – design built around the system, addressing issues such as: 'what can be built easily on a particular platform?'

Activity 2.1

1. Explain what is meant by the term 'human computer interaction'.
2. What factors should be taken into consideration in terms of human computer interaction?
3. What psychological factors could affect human computer interaction, and what steps can be taken to address and overcome these?
4. In what ways do different users impact on interface design?
5. What interface issues should be taken into consideration for expert user types?

The implications of developing and designing systems and software without reference to how end-users will interact with them are far reaching, and the impact of this can be detrimental for organizations. If users cannot interact effectively with a system or the software this could result in:

- low or no productivity
- delay in task requirements
- resentment towards system/software use
- refusal to undertake tasks that are specifically software driven
- further and frequent training (over and beyond any initial software training given).

Systems and software development revolves around a life cycle, one of the most important aspects of this life cycle being the interface design, i.e. how easy end-users will find using the software. The issue of HCI focuses on user-centred design, rather than systems-centred design.

User-centred design is based on a user's:

- working environment
- job role(s) and tasks
- abilities, needs and requirements,

the process of the design being collaborative between the end-user and the programmers and designers to ensure that user can interact effectively with the software and system.

For an organization a user-centred design approach can impact on certain resources and have the following implications:

- Costs may be increased as more user-friendly adjustments and customizations may have to be built into any software that is designed, especially for more specialist software.
- The development and implementation process may take longer as end-users would have input into the overall design.
- There may be conflicts of interest between what end-users want and what programmers want to develop.
- Issues of ownership may arise – if end-users have contributed to the development they may resent any future changes being made.

The need to customize software to develop a specialist HCI can be achieved through the use of specialized tools known as 'interface tools' and 'visual development tools'. These tools can focus specifically on the different stages of software development and they allow rapid graphical user (GUI) interface development.

Examples of these tools are:

- Microsoft Visual Basic
- Delphi
- Toolbook
- Java.

These tools provide interface capabilities through 'drag and drop' buttons, fields and combo boxes.

There needs to be a balance between functionality and how much HCI is considered in systems and software development and design. Any interface that is designed has to take into account a number of factors, some of which have already been explored. Interface design should also take into account the level of user (Table 2.1).

Table 2.1 User interface design profiles

User type	Interface type
Novice/first time: assumption that this user type has minimal knowledge of both task and interface	Simple
	Easily accessed
	Built-in help, tutorial facilities
	Graphical
	Drag-and-drop commands
Knowledgeable/intermittent user: some knowledge, may have a problem remembering functions and commands	Emphasis on recognition
	Consistent
	Context sensitive (to fill in knowledge gaps)
Expert/frequent user: familiar with task and interface, minimal prompts and reminders	Requires fast response
	Shortcuts available
	Textual

The need for a usable software interface is outlined in a technology article in *Computer Weekly* (Case Study 2.2). The article outlines the importance to end-users and an organization of designing and making the right software choice.

Case Study 2.2

Understanding the psychology of usability

No matter how well an IT system runs, if the user interface is wrong productivity will suffer. It is up to IT managers to put themselves in the users' shoes.

Many instruction leaflets supplied with flat-pack furniture can be nigh on impossible to follow. Why can't manufacturers make such products as simple to assemble and use as a child's toy?

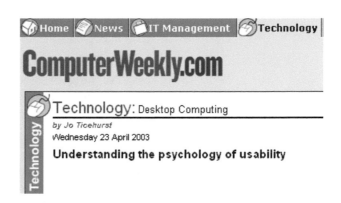

http://www.computerweekly.com/Articles/2003/04/16/193906/understanding-the-psychology-of-usability.htm.

The same can be true of software. As IT systems become increasingly complex, the learning curve for users is getting steeper. In some cases, making the right software choice can mean the difference between success and failure. There are many facets to user interface design that IT departments must consider before deciding what software to use. A well-designed, intuitive IT system means less training and support for end-users.

It is important that IT staff understand the end-user experience. 'You should put yourself in the position of your users,' said Robina Chatham, visiting fellow at the Cranfield School of Management. She warned that many IT professionals are drawn to technically challenging areas of functionality, but this does not necessarily produce easy-to-use software. 'Consider what your users really need rather than the features of the software. How much do they actually use?'

User-friendly interfaces

In any software, whether commercial or bespoke, giving the end-user too many options is confusing. 'If it is too complicated, many users will just give up.'

Brian Oakley, an expert on the subject of usability at the British Computer Society's Human-Computer Interface Specialist Group, agreed. 'Remember that the person is at the heart of the system. Many systems have gone wrong because people become enthralled by technology.' Oakley warned that most users hate delays. 'Three seconds is manageable, but when a system begins to take more than 20 seconds, users start doing something else,' he said.

Productivity is also affected if a system is too complex, or if end-users have been given inadequate training In how to use the software. Even something as simple as the colours used on your company's systems can affect how people work. 'People do not like to be bombarded with bright colours – their eyes can get very tired,' said Oakley.

He said application developers should avoid using garish hues such as bright reds and yellows and keep the number of colours to a minimum. Choose muted pastel shades such as green and blue when designing user interfaces, he said. Dresner feels that fantastically designed GUIs are not what users need if they are required to work in a particular way. He suggests designing the interface so that it is fit for purpose and satisfying to use. 'Think about the information the user will need on their screen,' he said.

So what exactly makes the definitive user interface? Microsoft is one company whose balance sheet relies on knowing what users want. It claims to have spent more than $3bn (£1.9bn) on the research and development of its Office productivity suite to build the most intuitive user experience in the product range.

One result of this research was a subtle change in screen colours. Daniel Bennie, Microsoft Office product marketing manager, said the research showed that users wanted text that was easier to read on screen. The result was that Microsoft introduced a grey border around e-mail messages in Outlook to make the message appear more prominent.

Besides colour changes, the formatting of text can have a huge impact on usability. Bennie said, 'We have also included a feature that enables users to read the text in columns like they would read a newspaper, as this is easier for the eye to scan.'

Activity 2.2

1. What are the key elements that IT staff and professionals should consider when developing software?
2. What does Brian Oakley warn about with regard to system delays?
3. Dresner comments in the article that you should: 'think about the information the user will need on their screen'. Why is this important?

Activity 2.3

1. What are the implications of designing software without reference to how end-users will interact?
2. Describe the concept of user-centred design.
3. There is a number of implications for an organization adopting a user-centred design approach. Evaluate two of these.
4. What is the difference between user- and systems-centred design?

HCI has grown and developed over the years as restrictions on hardware have been lifted and newer technologies have made tasks more flexible and manageable from both a developer's and an end-user's viewpoint.

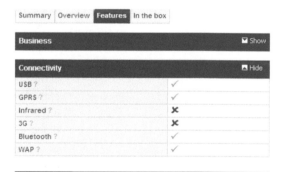

Figure 2.3 Example of how mobile technology has advanced.
http://www.carphonewarehouse.com/mobiles/mobile-phones/MOTOROLA-MOTOJEWEL/MONTHLY

Although there is quite a diverse range of systems available in terms of PCs, Macs and other devices, interfaces across this range seem to have evolved from text-based DOS-type systems and progressed to more graphical, interactive interfaces.

Advances in mobile technology and the need for good, functional, easy-to-use interfaces in devices such as mobile phones, laptops and personal digital assistants (PDAs), have given rise to interface features such as colour icon/menu-driven screens, voice recognition dialling, touch-sensitive displays and predictive texting. An example of this can be seen in the advances in mobile phone technology (Figure 2.3).

The days of command-line systems, which were an ideal environment for scientific and mathematical programming, have now been replaced with more visual systems and GUIs, and specialist software that can support various user types, for example those with visual impairments or special needs. In addition, 3D interfaces and voice recognition systems have emerged, fuelled by the rise in game consoles and games entertainment. The need for more realistic virtual reality systems that complement the needs of users has also become more prominent with the use of headsets, controllers and the more innovative Wii fit mat as shown in Figure 2.4.

Figure 2.4 Wii Fit board.
http://www.amazon.co.uk/Wii-Fit-Balance-Board-Games/dp/B001G0ATMM/ref=sr_1_2?ie=UTF8&s=electronics&qid=1234736857&sr=8-2

Activity 2.4

1. Different users require different features from an interface. Complete the table identifying an appropriate interface type for each user listed.

User type	Interface type
Novice/first time: assumption that this user type has minimal knowledge of both task and interface	
Knowledgeable/intermittent user: some knowledge, may have a problem remembering functions and commands	
Expert/frequent user: familiar with task and interface, minimal prompts and reminders	

Interfaces have changed over the past few years, the trend being towards more portable systems such as laptops, palmtops and mobile phone (WAP and Bluetooth) technology.

2. Provide a short summary of the benefits and limitations of the mobile phone interface (consider things such as use of colour, graphics and predictive text).

Society, the economy and culture

Improving usability of a device through the use of good HCI design as previously explored is paramount to the development and success of any system or piece of software.

User interfaces, although quite standard in terms of being used across similar technologies such as mobile phones or games consoles with similar features and interaction tools, can be quite diverse in certain specialist environments. Examples of these are environments for users who may have a special need due to a disability with speech or sound, or interfaces for hostile environments, for example 'fly by wire'.

HCI developments and progress made in interface technology have had an economical impact on certain types of users, especially users working within organizations. HCI can impact on:

- productivity levels – speeding up inputs
- reducing the complexity of inputs and tasks
- improving and increasing productivity levels
- enhancing the working environment by using different data capture, input and output methods.

The way in which people interact with systems and the levels of interaction with systems such as computers, mobile technologies, portable devices and games consoles have changed and increased over the years. HCI has contributed to this growth through the excellent interactive, innovative and user-friendly screen designs and interfaces that have made interaction with certain devices quite addictive.

From a psychological and sociological viewpoint HCI has contributed to a possible deskilling of the workforce as tasks become more approachable and easier to manage owing to the systems that are used.

The gulf between technologies and the influence of good HCI designs in systems such as mobile phones, laptops and games consoles has further contributed to the divide between developed nations that have access to these technologies and developing nations that do not.

Understand the fundamental principles of interface design

Good interface design is essential if the system for which it is designed is to be embraced by end-users. Poor interface design can lead to systems becoming too difficult to engage with, thus leading to non-usage and possible rejection by a user or an organization.

Perception

Perception plays a major role in HCI and interface design. The way in which users interact and view a particular interface can be the deciding factor between the success or failure of a particular system or device. Perception can be broken down into a number of areas, as shown in Figure 2.5.

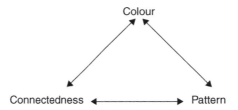

Figure 2.5 Perception elements in interface design.

Colour is extremely important in terms of interface design and can have a major impact on how users engage with a particular system or device. Certain colours on certain backgrounds can be quite difficult to view, and some users may be colour-blind or have a visual impairment that could make it more difficult to differentiate between colours. All of these issues should be considered when incorporating colour into any interface design.

Colour can be characterized in terms of:

- **a trichromatic system** – based on the primary colours and how these are perceived by the brain
- **luminance** – the amount of light that is reflected or omitted
- **popout effect** – the use of colour to make images and characters stand out.

The pattern or design on an interface is an area that in conjunction with colour can be quite problematic if time has not been invested in researching designs or the wallpaper/interface background.

Pattern could take into consideration the proximity of objects or graphics – how close are they to another object or graphic? Does the interface look too cluttered or sparse? The continuity and symmetry of a pattern could aid recognition if users are used to certain items being placed in certain areas; if there is continuity in terms of the use of certain icons or menu functions then this could prove to be advantageous. A similarity to known interfaces or other familiar objects, tools or graphics and the common grouping of these can also be beneficial, especially to novice users.

Behaviour models

A number of behaviour models can be used to help understand the fundamental principles of good interface design. These models can be divided into two categories:

- **predictive models** – reaction time, Keystroke Level Model (KLM), Throughput (TP), Fitts' Law
- **descriptive models** – Key-Action Model (KAM), Buxton's three-state model, Guiard's Model.

Keystroke Level Model

This model is based on the time it takes for a user to complete a task on a computer. The prediction is based on a number of variables such as:

- how many tasks have to be completed/accomplished
- what methods are being used to complete the tasks
- response times
- the motor skills of the user
- the command language of the computer.

Throughput

This is also referred to as 'index of performance' or 'bandwidth' in Fitts' Law tasks is a metric in quantifying input system performance.

Fitts' Law

Fitts' Law is an ISO standard for evaluating computing pointing devices. The model is based on time and distance. It predicts time for rapid aimed movements based on the target of a specified size at a specified distance and establishes performance differences between devices and interaction techniques.

Key Action Model

This model examines how a user will expect a computer to respond or behave under certain conditions and how this might differ from how a computer actually behaves.

Buxton's Three-State Model

This model examines movements that a user makes when interacting with a mouse or a keyboard. At the user interface design stage, consideration should be given to the amount of pressure that a user will need to apply in order to get a command response. In conjunction consideration should also be given to the ease of use.

Guiard's Model

Humans are not only two-handed, they use their hands differently; this model examines a users preferred method of interacting with computers and input devices. Hands have different roles and perform distinctly different tasks. With the knowledge that people are either right-handed or left-handed, Guiard's model examines the action of preferred and non-preferred hands for given tasks and roles.

Information processing

When designing any user interface, one of the most important elements that should be taken into consideration is the human element – what does the end-user require and expect from the interface?

In terms of the information processing element there are two aspects that should be considered, the 'human information processing' (HIP) element and the 'goals, operators, methods and selection' (GOMS) element.

Be able to design and produce simple interactive input and output based on HCI principles

Designing systems and devices that have simple interactive input and output elements can be quite challenging, especially as different users have different ways and expectations of capturing, inputting, processing, storing and outputting information.

Input

There are several traditional input devices such as a keyboard and a mouse that are associated with devices such as computers, laptops, PDAs and other mobile technologies. In addition to these there are input devices that are associated with certain technologies and devices such as joysticks, joypads, trackerballs and more specialist or niche devices such as concept keyboards, nun chucks, wands and console mats or boards.

Some input devices are quite straightforward and rely on the user simply typing in a set of commands or instructions using a keyboard or keypad, such as on a computer. Some of these can be touch sensitive, for example on a range of mobile phones such as the LG Prada or iPhone. Predictive texting on mobile phones is another way of supporting the end-user by minimizing the keystrokes needed to compose a message.

Some input devices are driven by magnetic strips in cards that may be used to secure or unlock devices, and some can be voice activated or driven by other means of recognition such as a fingerprint or iris scanning.

Output

The way in which information is displayed or output can also vary depending on the device or system. The majority of devices are dependent on the use of a screen or monitor to output information, which can vary from a traditional computer monitor through to a mobile phone display, PDA screen or satellite navigation panel, for example.

With a monitor or screen, the important design elements to consider include the size, use of colour, clarity and layout.

Information can also be output straight to a printer, or projected onto a screen or an interactive whiteboard.

Specialist

Many people require specialist input or output devices to interact with ICT.

The range of ICT available is extensive, and various organizations are committed to providing these devices to users who may have a sensory impairment, limited mobility or a language disability. In addition, certain people may have a particular need as a result of an illness resulting from an accident, an inherited illness or a general learning disability. These conditions can affect their motor control or dexterity, or they may have limited use of limbs or a short attention span.

Some of these impairments, conditions and disabilities will be explored in this section, with references to organizations that have developed and are continuing to develop technologies to enable users with specialist needs to interact with ICT.

Technologies for the Visually Impaired Inc. (http://www.tvi-web.com) offer a wide range of adaptive devices, software and accessories specifically designed for use by blind or visually impaired individuals. The range includes adaptive computers, reading machines, Windows access software, speech synthesizers, refreshable Braille products, voice recognition software, screen magnification software, CCTV products, Braille embossers, Braille translation software, tactile imaging products, various accessories, customized personal computer systems, and much more.

The Royal National Institute for the Deaf (Case Study 2.3) and more specifically the New Technologies team have made huge advances in the area of facilitating deaf people, and enabling them to communicate and embrace ICT.

Case Study 2.3

The RNID New Technologies team

Today's 'information society' has brought many benefits, but at the same time it has created new problems for those people who have particular abilities and preferences for using information and communication technology.

The New Technologies department at RNID has two **overall objectives**:

1. To harness information and communication technology to bring tangible benefits to deaf and hard of hearing people and to remove barriers they face in society.
2. To participate actively in the various technology communities so that both existing as well as emerging technologies become more inclusive for deaf, hard of hearing and speech-impaired people. The areas in which we are active, include:

   - The internet
   - The World Wide Web
   - Mobile and wireless networking
   - Digital broadcasting
   - Information technology

Projects

Communicating while on the move

While a couple of decades ago, mobile phones were a luxury gadget used by a small sector of society, the GSM (digital mobile phones) revolution has dramatically changed that. These days, being able to communicate while on the move has become absolutely essential.

However, quite a number of deaf and hard of hearing people are unable to communicate in this manner and face increasing barriers in employment, education, social life and entertainment as a result. We have developed solutions for mobile text telephony and are continuously improving on these. The problems for sign language users and access to personal information on mobile handsets for all deaf and hard of hearing people are also being addressed.

We work very closely with handset manufacturers, network operators and service providers to make them aware of the requirements of deaf, hard of hearing and speech impaired users.

One of the solutions developed at RNID is the mobile textphone based on the Nokia 9210 Communicator.

The case for interactive texting

"Many deaf and hard of hearing people cannot use their voice to communicate, and use real-time text instead, which comes in the form of interactive, character-by-character texting.

Unfortunately, traditional methods to deliver real-time text (such as textphones) have been non-mainstream solutions: expensive, not very user friendly and with limited functionality. However, using text to communicate is not an exclusive activity for deaf people. Today, almost everyone uses SMS, e-mail and even 'instant messaging'.

We are working very hard to integrate the needs of deaf, hard of hearing and speech-impaired people into the mainstream offering. If we get it right, deaf and hard of hearing people will be able to use text to communicate with others anywhere and anytime."

European projects

Links to international projects, funded by the European Commission, also allow the department to work together with experts across Europe and to drive key research in areas such as speech recognition, signing avatars (virtual humans) and web site accessibility.

Some of our European projects include:

- VisiCAST
- eSign
- Synface

Source: RNID.org.uk

There is a wide range of devices available to support individuals with special needs (Table 2.2). For example:

- speech amplifiers and induction loops
- speech synthesizers
- text telephones.

Table 2.2 Overview of devices available to support specialist users to interact with ICT

Condition or disability	Devices available as described in the chapter
Sensory impairment	Reading machines, Braille products, videophones
Limited mobility	Adaptive computers, videophones
Language disability	Speech synthesizers, text processors, e.g. Wordcat
Lack of motor control or dexterity	Adaptive computers
Limited use of limbs	Headsets, text phones and communicators
Short attention span	Signing avatars (virtual humans)

A wide range of hearing aids supports induction loop technology, which has been used for decades in banks, libraries, hotels and other public premises to help deaf people to communicate. The hearing aid captures the signal supplied from the loop when it is in the T-mode.

Videophone

In the multimedia project of the Finnish Association of the Deaf, new forms of services have been created with the help of modern technology. In the Joensuu region the videophone is used for long-distance interpreting in the Finnish Sign Language and in the northern part of Finland for bringing social services within the reach of deaf people.

A deaf person together with a hearing person can be seated in front of the same videophone and together make picture contact with the Joensuu interpreter referral service centre. Alternatively, a deaf person can use the videophone to contact a Finnish Sign Language interpreter who connects him/her to a hearing person with a regular phone and vice versa.

Communication software

VTT Information Technology has developed WordCat software. WordCat is a text processor input device for people with physical and speech disabilities. Text can be created through individualized wordlists and phrase libraries as shown in Figure 2.6.

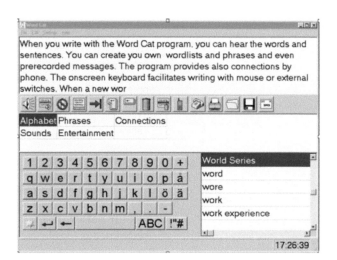

Figure 2.6 WordCat screen.
http://www.RNID.org.uk

Speech synthesizers

Speech synthesis has developed steadily over the past decade and it has been used in several new applications. For most applications, synthetic speech has reached an acceptable level. However, in other areas such as text processing and pronunciation there is still a great deal of work and development to be done to ensure that any speech sounds more natural. Natural speech is so dynamic, however, that replicating it may be impossible to achieve.

Text telephones

People with hearing or speech problems may use a text telephone. Unlike a standard telephone, a text phone has a keyboard and a display screen. Instead of speaking into a telephone mouthpiece, information is typed in, using the keyboard.

In the UK, the word 'Minicom' is often used to describe any text phone. Minicom is a widely used brand of text phone. When a text phone is used to call someone with a voice telephone, when they pick up the telephone, they will hear a prerecorded message such as 'hard of hearing caller, please use a text phone', or they will just get silence. If they also have a text phone, they should be able to take the call and continue with the conversation. If they do not have a text phone, then a 'relay service' will need to be used to call them back. At present, the only national relay service in the UK is RNID Typetalk.

Be able to compare and contrast, using HCI principles, the effectiveness of different designs of input and output

The effectiveness of different input and output designs can be measured in various ways. These measurements can be based on quantitative or qualitative effectiveness, examine elements such as speed, costs and user satisfaction, and undertake comparative analysis with other systems and designs.

Quantitative and qualitative measures of effectiveness

Quantitative measures of effectiveness can be based on speed, costs and comparisons made with the original needs in terms of the features available and in comparison with other systems.

Speed can be quantified in terms of:

- **Input speeds** – how long does it take to interact and physically input the necessary data or information?
- **Keying speeds** – does the input device provide optimum opportunities for the user to key in information quickly and with precision?
- **Throughput** – the time it takes to fully engage and input the required data or information in preparation for the next stage of processing/ storage or output.
- **Comprehension of output** – how long does it take for information to be recognized that is ready to be output by a preferred method? For example, is the report on the latest sales figures in a suitable format ready to be printed?

Costs are also a quantitative measure of effectiveness in terms of design costs, running costs and staffing costs. How much will the initial design cost? How much will it cost to maintain? Will there need to be any investment in training new users in the system, or will more staff need to be employed to monitor, use or maintain it?

Comparing the system with the original requirements or end-users' needs is also a good quantitative measure of effectiveness. Have all of the original features been included? If not, how many or why not? How does the system compare with other systems, etc?

Qualitative measures of effectiveness can be based on the overall experience of use. How satisfied was the end-user? How easy was the device or system to use? Were the results or was the output as expected? Did it meet the end-users' needs?

Comparisons can also be made with other systems to ascertain whether the newer system is more efficient, easier and more cost effective, for example, than other systems, and in what way.

Evaluation of interfaces

The evaluation of an interface can be based on both quantitative and qualitative judgements. Judgements based on its effectiveness can also be made in terms of how closely the new interface meets the original requirements, what the good and bad points of the system are, and what improvements can be or need to be made.

Activity 2.5

1. Look at a range of interfaces for a particular system, for example laptops, mobile phones, PDAs or satellite navigation systems.
2. Complete a table similar to the one provided that compares each interface based on a range of qualitative and quantitative criteria.

Device	Cost	Speed of input	Ease of use	Overall user experience rating (1–10)[a]
1				
2				
3				

[a]1 = lowest; 10 = highest.

3. Include pictures or screenshots of the three interfaces that you choose for your device or system.

Questions and review

1. What impact can HCI have on society, the economy and culture?
2. In terms of HCI developments, how and in what ways were early designs restricted?
3. Interfaces for visually impaired people are one example of a specialized interface that is currently used in today's society. Provide two other examples of specialized interfaces.
4. What is meant by 'fly by wire' technology?
5. How can HCI impact directly on productivity within an economical environment?
6. Culture can have a direct impact on HCI design. Can you provide an example where this has happened?
7. There are several fundamental principles associated with interface design. How can perception impact on these?
8. What is meant by a behaviour model? Provide two examples of a behaviour model.
9. What does 'human information processing' (HIP) mean?
10. Identify three ways in which information can be input into a system.
11. What output considerations should be addressed in terms of HCI design?

12. What considerations and adaptation may have to be made to an HCI design if it is aimed at a specialist audience?
13. Identify three quantitative measures of effectiveness in terms of HCI input and output design.
14. Identify three qualitative measures of effectiveness in terms of HCI input and output design.
15. In what ways can the evaluation of interfaces take place?

Assessment activities

Grading criteria	Content	Suggested activity
Pass		
P1	Describe one impact of HCI in recent years on each of society, economy and culture.	Carry out research to look at how HCI has impacted upon society, the economy and culture over recent years. Produce an information leaflet.
P2	Explain two fundamental principles of HCI design.	Produce a short presentation that explains two fundamental principles of HCI design.
P3	Design and create three input HCI to meet given specifications, using a variety of techniques.	Design and create three input and three output HCI to meet given specifications, using a variety of techniques.
P4	Design and create three output HCI to meet given specifications, using a variety of techniques.	Design and create three input and three output HCI to meet given specifications, using a variety of techniques.
P5	Briefly describe how each of the input and output HCI they have created meet the specifications provided.	Add a brief written extract to the leaflet that describes how each of the input and output HCI created meet the specifications provided.
Merit		
M1	Explain how modern advances in HCI design have contributed to the impact of computers on society, economy and culture.	In conjunction with P1 and D1, extend the leaflet to explain how modern advances in HCI design have contributed to the impact of computers on society, economy and culture.
M2	Design one HCI dedicated to specialist needs indicating how some of the fundamental principles have been applied and how the needs are met.	In conjunction with P3 and P4, develop the input and output screens into a HCI that is dedicated to specialist needs. In conjunction with the design an explanation should also be provided as to how some of the fundamental principles have been applied and how the needs are met.
M3	Describe how effectiveness of HCI may be measured.	Produce a short presentation describing how the effectiveness of HCI can be measured.
Distinction		
D1	Evaluate the HCI developments over recent years, relating them to the impact on society, economy and culture, and predicting one potential future development and what impact that may have.	In conjunction with P1 and M1, a short evaluation could be produced, linked into the leaflet that evaluates HCI developments over recent years and relate them to the impact on society, economy and culture. You should also predict one potential future development and what impact that may have.
D2	Compare the HCI they have developed with those commercially produced for similar products, indicating the good and less good features of each and any improvements which could be made.	In conjunction with P3, P4 and M2, compare your own HCI with those that are commercially produced for similar products. Identify a range of features for each and any improvement that could be made.

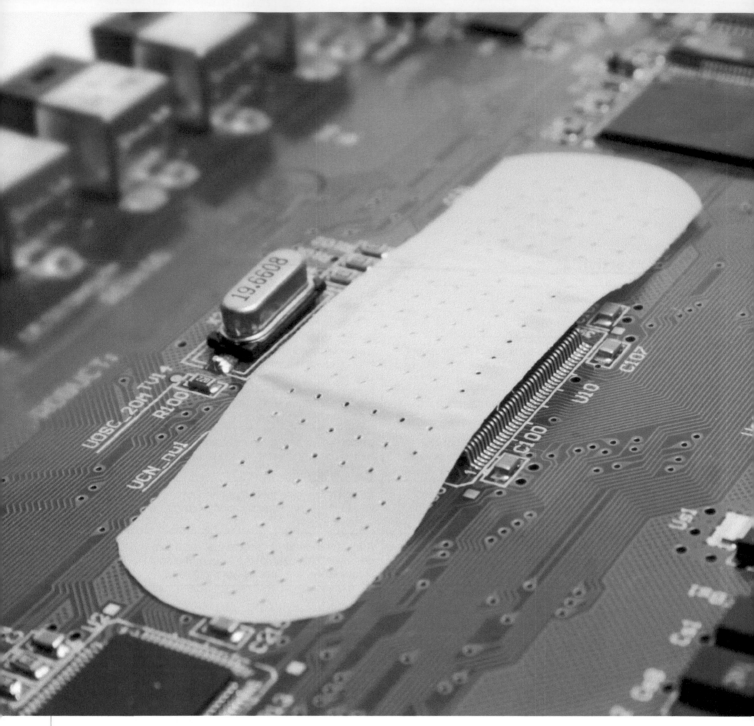

The maintenance of computer systems is an essential part of ensuring that hardware, software and communications are functioning effectively and efficiently.

Chapter 3

Maintaining Computer Systems

I n today's society there is a growing need for skilled practitioners who have the knowledge and application to maintain systems at various levels of complexity.

Larger organizations will have IT departments with a number of specialists ranging from helpdesk support through to designers, programmers, hardware maintenance, network specialists and IT managers. Therefore the roles of individuals are quite defined and possibly restricted to a certain area. In smaller organizations there may just be one IT person who has a number of roles and a wider set of responsibilities.

Computer maintenance is an essential part of any IT role. It is essential that problems can be diagnosed and addressed accurately. This chapter will provide you with information and resources about how to plan and perform computer system maintenance. In addition you will be introduced to the health and safety aspects of maintaining computer systems.

There is an expectation that you will be able to perform routine housekeeping on computer systems and monitor these systems.

The chapter is structured around the following learning outcomes:

- Understand the importance of planning and performing computer system maintenance.
- Know health and safety risks and precautions associated with maintaining computer systems.
- Be able to perform routine housekeeping on computer systems.
- Be able to monitor systems and perform necessary enhancements to systems for improvements in performance and efficiency.

Understand the importance of planning and performing computer system maintenance

Computer system maintenance can be proactive in terms of planning for updates and new system installations, or it can be reactive in terms of responding to an immediate fault or issue.

Maintenance

The scope of computer maintenance can extend from very routine and simplistic tasks such as cleaning the equipment and replacing cartridges in printers, to more complex and non-routine tasks such as resolving network conflicts and issues of platform incompatibility between systems.

Routine maintenance tasks may be carried out at set periods, for example every week, whereas non-routine maintenance may occur on demand when a problem arises.

Activity 3.1

Both routine and non-routine tasks can be carried out on hardware or software.

In the table below is an example of both routine and non-routine maintenance tasks that could be carried out on a printer.

Printer	
Routine tasks	**Non-routine tasks**
Clearing the print queue	Changing the fusing unit on a laser printer
Replacing the paper	Replacing components
Sorting out a paper jam	
Changing the cartridge	

1. Using the example above, create a table that identifies a range of routine and non-routine tasks that can be carried out on a component or system.

If maintenance tasks are not carried out, this could result in a delay of work tasks, increased costs and possibly system failure, which may result in the system being replaced rather than repaired.

Maintenance tasks may be carried out in-house by the IT department, or the maintenance can be outsourced to a third party. A range of maintenance specialists provide onsite support or 'back to base' support, where items are returned for repairs rather than maintenance being carried out on the premises.

In addition, some organizations will opt for remote support and maintenance, where a fee is paid to a company which can then monitor and conduct maintenance over the Internet. An example of this is the service offered by Beam2Support, as shown in Case Study 3.1.

Case Study 3.1

Beam2Support

Beam2Support

Case Study – Yamaha Music Central Europe GmbH

YAMAHA

YAMAHA is one of the world-leading manufacturers of music instruments and equipment. For over 40 years the company has been active in Europe. In 2001, YAMAHA Music Central Europe generated revenues of over €200 Million. Currently they have over 600 employees in Europe and operate over 600 music schools.

"Beam2Support completely convinced us because of the easy usability for both the supporter as well as the field agent who is looking for help."

Heiko Harder
IT-Division / Software Engineer
YAMAHA Music Central Europe
GmbH

The Challenge

The sales area of the YAMAHA Music Central Europe GmbH covers the entirety of Europe. Within Europe alone, they have over 70 outside field agents. Therefore, the only way to support these agents was by telephone. Heiko Harder, Software Engineer of YAMAHA Music Central Europe GmbH greatly needed a solution to optimize the support time for the field agents and the support team. *"We needed a remote support solution to support our agents in the field with their hardware and software problems."*

The Solution

With the implementation of Beam2Support, the support staff members were able to connect to the field agents' PCs without preparation. They could then immediately diagnose and solve difficult support cases. *"With the help of the screen sharing, we are able to detect the problems much faster than we were previously with just the telephone."* The highly user-friendly solution of Beam2Support offers a connection within only a few seconds. *"The incredible ease of use of Beam2Support convinced us immediately."*

The Conclusion

With Beam2Support YAMAHA was able to more effectively and economically support their field agents. Implementing system updates and resolving daily support cases were able to be done more efficiently. *"So far we have had only positive feedback from our agents. Beam2Support is a very cost effective solution and we are happy to recommend it as a support tool."*

http://www.beam2support.com/EN/customers/casestudiesjs.aspx

Activity 3.2

Read through the case study on the service provided by Beam2Support.

1. What are the benefits offered to Yamaha through this remote service?

Organizational policy and procedure

Organizations will have a number of policies and procedures in place that provide the framework for areas such as:

- procurement – how organizations buy new hardware and software
- sustainability and environmental issues
- reporting procedures

CHAPTER 3

- documentation and problem escalation procedures
- employee responsibilities, conduct, etc.

Procurement ensures that organizations acquire the right hardware and software, goods and services at the right price, in the right quantities and to meet the right conditions. Procurement may involve putting tenders out to suppliers to ensure that these conditions are met.

There are several factors to consider when entering into a procurement process, including:

- gathering information about the conditions, needs and requirements of the hardware/software acquisition
- selecting a supplier or preparing a tender to supply
- analysing the tenders and other supporting information regarding the purchase
- negotiating and awarding contracts
- fulfilment – delivery, implementation, payment and training
- evaluation and review of the process
- renewal and update of the process and planning for the next procurement activity.

E-procurement follows a similar life cycle of stages; however, transactions and tenders are sent out electronically through the Internet and using technologies such as electronic data interchange (EDI).

Organizations today are becoming more aware of sustainability and environmental issues. The drive towards a 'greener' environment has influenced the policies and procedures of a range of organizations at all levels.

Case Study 3.2, on Dell, demonstrates how organizations can take steps to address issues of sustainability and environmental issues.

Case Study 3.2

Dell Recycling

Dell is committed to becoming the greenest technology company on the planet. We are reducing our environmental impact by designing energy-efficient products, undertaking responsible manufacturing and running easy-to-use recycling programs. Environmental considerations are built into every stage of the Dell product life cycle. Our measures start with development and design, span manufacturing and operations and continue across customer use and product recovery.

Other environmental measures at Dell include increasing our use of recycled-content paper and paper certified by the US Forestry Stewardship Council.

One of the most direct ways we can reduce our impact on the environment is by helping our customers recycle their computer equipment – either to make the most of devices that still work, or to dispose of old equipment responsibly. By doing this, we help prevent waste and pollution.

Dell is a global industry leader in recycling. In 2006, we became the first technology company to offer free recycling of its products to consumers anywhere in the world – with no exceptions. In 2006, our product recovery programs nearly tripled – growing 264 percent. Dell is the first computer company to offer no-charge worldwide computer recycling for its products. As a result, we dramatically expanded our consumer recycling programs. Dell's goal is to recover 125 million kilograms of discarded product by fiscal year 2010 through asset recovery programs. Since introducing its computer recycling service to Australia and New Zealand in 2004, Dell Australia has helped customers reuse and recycle more than 800 tonnes of computer equipment.

http://supportapj.dell.com/support/topics/topic.aspx/ap/shared/support/recycle/en/recycle?c=au&l=en&s=genment.

Dell's Environmental Policy

Dell's vision is to create a company culture where environmental excellence is second nature. Our mission is to fully integrate environmental stewardship into the business of providing quality products, best-in-class services, and the best customer experience at the best value. We have established the following environmental policy objectives to achieve our vision and mission.

Design Products with the Environment in Mind

- Design products with a focus on: safe operation throughout the entire product life cycle, extending product life span, reducing energy consumption, avoiding environmentally sensitive materials, promoting dematerialization, and using parts that are capable of being recycled at the highest level.
- Set expectations of environmental excellence throughout Dell's supply chain.

Prevent Waste and Pollution

- Operate Dell's facilities to minimize harmful impacts on the environment.
- Place a high priority on waste minimization, recycling and reuse programs, and pollution prevention.

Continually Improve Our Performance

- Use an Environmental Management System approach to establish goals, implement programs, monitor technology and environmental management practices, evaluate progress, and continually improve environmental performance.
- Foster a culture of environmental responsibility among employees and management.

Demonstrate Responsibility to Stakeholders

- Act in an environmentally responsible manner through sustainable practices designed to ensure the health and safety of Dell's employees, neighbours, and the environment.
- Periodically communicate company progress to stakeholders.
- Engage stakeholders to improve products and processes.

CHAPTER 3

Comply with the Law

● Conduct business with integrity and dedicated observance of environmental laws and regulations, and meet the commitments of the voluntary environmental programs in which Dell participates.

http://www.dell.com/environment

Activity 3.3

1. What steps have Dell taken to address issues of sustainability and environmental issues?

2. Can you find a similar case study of an organization within the IT industry that has addressed similar issues?

Reporting on the process or steps taken in any computer maintenance activity is vital to ensure that there is a record of when and what occurred. This record is needed in case a problem happens in the future, when documentary evidence can be used as an audit trail to eliminate faults. In addition, through the reporting system, records of when hardware and software were replaced, installed or configured can be logged for future procurement activities.

In conjunction with the reporting system a range of documentation should be retained. This could include manuals and support guides, warranties, hardware and software audit documentation and fault logs that provide a historical account of the problems that have occurred, and escalation procedures. Documentation should also be retained regarding the roles and responsibilities of people within the department. This defining of roles will ensure that the right technical expertise is directed to the right technical issue. In addition, boundaries can be established in terms of end-user support, design, development and maintenance tasks.

Planning

A number of planning techniques and documents can be used to support a range of computer system maintenance activities and other IT-related tasks. These include:

● route maps
● upgrade paths
● schedules
● Gantt charts
● operational planning.

Activities can be monitored and planned using tools and techniques such as route maps, upgrade paths and schedules. Activities may include the implementation of hardware, installation or upgrade of software, or the complete overhaul of a networked system.

Gantt charts can be used to map and schedule specific tasks, as shown in Figure 3.1. They can illustrate how long a particular task or project will take and also identify task dependencies. This can be critical in events such as upgrading systems where the timing of hardware and software implementation is crucial.

Task	possible start	Length	Type	Dependent on...
1. High level analysis	week 1	5 days	sequential	
2. Selection of hardware platform	week 1	1 day	sequential	1
3. Installation and commissioning of hardware	week 3	2 weeks	parallel	2
4. Detailed analysis of core modules	week 1	2 weeks	sequential	1
5. Detailed analysis of supporting utilities	week 1	2 weeks	sequential	4
6. Programming of core modules	week 4	3 weeks	sequential	4
7. Programming of supporting modules	week 4	3 weeks	sequential	5
8. Quality assurance of core modules	week 5	1 week	sequential	6
9. Quality assurance of supporting modules	week 5	1 week	sequential	7
10. Core module training	week 7	1 day	parallel	6
11. Development of accounting reporting	week 6	1 week	parallel	5
12. Development of management reporting	week 6	1 week	parallel	5
13. Development of management analysis	week 6	2 weeks	sequential	5
14. Detailed training	week 7	1 week	sequential	1–13
15. Documentation	week 4	2 weeks	parallel	13

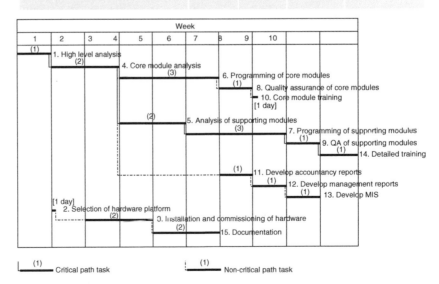

Figure 3.1 Gantt chart example. Mindtools: http://www.mindtools.com/pages/article/new PPM_03.htm.

Operational planning will involve decisions that need to be taken on a day-to-day basis. Planning may include upgrading the network to ensure that it can accommodate an increase in traffic or designing a website to market the services of the organization.

Know health and safety risk and precautions associated with maintaining computer systems

Legislation

In conjunction with the Health and Safety at Work Act (1974) additional European legislation was passed in 1992. This legislation has become known as the 'Six Pack' and it sets a minimum standard of provision for health and safety across the European Community. It incorporates maintaining good working practices and having positive and actionable health and safety policies.

The current regulations focus on the basic requirement to undertake risk assessments that will identify hazards. These hazards, if they cannot be eliminated, should be assessed to establish the degree of risk to which an employee would be exposed while at work, and train that employee to work safely with these identified risks.

In terms of ergonomics, the following considerations should be made in terms of:

- hardware
- workstation
- software
- furniture and its layout.

The workstation should be designed to provide sufficient space for the user to change position and vary movements. Users' daily work activities should also involve other tasks other than just using the visual display unit (VDU). It is also essential that people who frequently use a VDU take regular breaks. In terms of the VDU screen, screen images should be stable and flicker free. The top of the VDU should be just below eye level and filters should be made available to reduce glare and reflection.

One health and safety act that has been set up to protect users of information systems and general ICT is the Display Screen Equipment (VDU) Regulations 1992.

Display screen equipment (VDU) regulations 1992

Under these regulations an employer has six main obligations to fulfil. For each user and operator working in his undertaking, the employer must:

(i)   assess the risks arising from their use of display screen workstations and take steps to reduce any risks identified to the 'lowest extent reasonably practicable'

(ii) ensure that new workstations ('first put into service after 1st January 1993') meet minimum ergonomics standards set out in a schedule to the Regulations. Existing workstations have a further four years to meet the minimum requirements, provided that they are not posing a risk to their users.

(iii) inform users about the results of the assessments, the actions the employer is taking and the users' entitlements under the Regulations.

For each user, whether working for him or another employer (but not each operator)

  (iv) plan display screen work to provide regular breaks or changes of activity.

In addition, for his own employees who are users:

  (v) offer eye tests before display screen use, at regular intervals and if they are experiencing visual problems. If the tests show that they are necessary and normal glasses cannot be used, then special glasses must be provided.

  (vi) provide appropriate health and safety training for users before display screen use or whenever the workstation is 'substantially modified'.

In terms of other hardware and workstation items, the keyboard should be placed in a position where the user has room to manoeuvre it comfortably and at a height where the forearms, hands and wrists can be in a straight line when typing. Ergonomically designed keyboards are available on the market that can relieve pressure on the wrists, arms, neck and shoulders. Wrist and arm supports are also available to relieve the pressure, when typing.

Ergonomic software can be classified as measures and steps that are taken to ensure that an end-user can interact with software effectively as a result of acceptable colours, graphics, font size and other human computer interface (HCI) issues. However, ergonomic software can also refer to the range of software that is available to support organizations in terms of 'risk assessment'. Software is available that will look at the health and safety procedures in place and assess how ergonomically orientated your organization is.

In terms of furniture and its layout, people generally underestimate the importance of the office chair. The Workplace (Health, Safety and Welfare) Regulations 1992 includes seating as well as workstations. The right office chair used correctly is fundamental to the welfare of the users.

It is important to ensure that seating in the workplace is safe, suitable for the task performed and comfortable.

- Employers are required to provide seating that is safe and meets the needs of the individual. It should also be suitable for the tasks involved.
- The work chair should also be stable and allow the user freedom of movement and a comfortable position.

Consideration should also be given to the positioning of wires, filing cabinets, natural lighting and adequate ventilation. The layout of a room can have a profound effect on an employee, therefore consideration and adequate planning should be undertaken before any workstations are installed so that provisions can be made for cabling, routers, servers, etc. In addition, adequate space needs to be provided so that users are not crammed into small office areas nor restricted by items such as shelving, filing cabinets and desks.

Health and safety risks

Precautions

A number of initial steps and precautions should be taken when installing hardware, including:

- shutting down and switching off the computer
- unplugging the power cord from the wall socket or rear of the computer
- giving yourself sufficient room to move around the computer and desk area.

Computer chips and hardware, such as motherboards and hardware cards, are sensitive to static electricity. Before handling any hardware or working on the inside of the computer you should also ensure that you have discharged the static electricity from your body. At the very least you should make sure you unplug your computer from the mains and touch the bare metal case with your hand to discharge any static that may have built up in your body. Ideally, you should wear a grounding strap and/or use an antistatic mat to reduce the risk of any components being 'zapped' by static.

Consideration should be given to the working environment of users of ICT. Issues exist which cover environmental, social and practical aspects of working conditions (Figure 3.2).

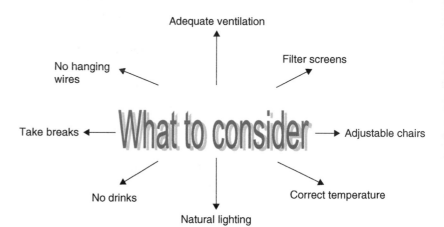

Figure 3.2 Health and safety considerations.

Users should be working in an environment that has adequate ventilation and natural lighting, and the temperature should be conducive to a computing environment, especially as computers give out large quantities of heat.

Computer users should also have sufficient support peripherals such as filter screens to minimize glare and height-adjustable chairs. When working at a computer no food or drink should be consumed in case liquid or crumbs fall onto the keyboard or into the case. Wires should always be packed away in appropriate conduits and not left trailing across the floor.

The best measure for health and safety in the workplace is to use common sense and adhere to standard ways of working. Organizations also offer guidelines and procedures for maintaining good working practice.

The need for such legislation can only benefit users and help to protect them against computer-related injuries such as:

- repetitive strain injury (RSI)
- back and upper joint problems
- eye strain
- exposure to radiation and hardware ozone
- epilepsy
- stress-related illnesses.

The Electricity at Work Regulations (1989) requires the maintenance of fixed electrical installations and portable devices to be carried out and to ensure that regular inspections take place to ensure that they are safe. Electrical test certificates should be issued to validate the inspection process.

Under UK law employers are responsible for ensuring the safety and welfare of their employees and also the public. In terms of electrical safety at work, a number of health and safety officers have been appointed across a range of disciplines to ensure that standards are maintained.

Activity 3.4

A range of information and guidance has been provided by the Health and Safety Executive (HSE) for the following areas:

- safety in electrical testing
- servicing and repair of audio
- television and computer equipment
- maintaining portable electrical equipment in offices and other low-risk environments.

1. Visit the HSE website: http://www.hse.gov.uk/ and look at the leaflets and information sheets that cover these areas, for example: http://www.hse.gov.uk/pubns/indg354.pdf (Safety in Electrical Testing).
2. Prepare an information leaflet aimed at new employees that covers each of the bullet points listed. The information should be based on your findings on the HSE website and other research sources if appropriate.

There are various ways in which health and safety risks can be minimized when working with computer systems and electrical supplies. Precautions must be taken to reduce the risk of fire, electrocution and electrostatic discharge (ESD). Some of these precautions will be based on the overall layout and ergonomics of the environment in which you are working. Other precautionary methods may involve the use of specialist equipment such as ESD wrist-straps or mats.

The most effective way of avoiding ESD is to use a wrist-strap or grounding mat. Other measures and precautions that can be taken to avoid ESD include:

- Remove any jewellery.
- Try to avoid wearing any clothing that may conduct ESD, such as woollen items.

- Remove any cords away from the back of the computer.
- Avoid working with computers in extreme weather conditions such as during an electrical storm.
- Continuously touch unpainted metal surfaces to ensure that you and the computer are at zero potential.
- Try standing rather than sitting, especially because the chair can generate more electrostatic charge.

In addition, fire-fighting equipment, canisters and trained individuals who can manage a potentially hazardous situation are required when working under conditions where a threat or risk may be present.

Be able to perform routine housekeeping on computer systems

Carrying out a range of housekeeping tasks is an important element in terms of maintaining computer systems. Ensuring that components are functioning efficiently by managing, cleaning, updating and maintaining them is essential on a day-to-day or periodic basis.

Managing file systems

Managing file systems is important because ensuring that data has been stored securely, in the correct format and in an easily accessible area can save time and money.

Organizing and applying naming conventions to files and folders is one of the most common tasks that is undertaken by everyday users working with data. Saving and naming a file is quite a simple process, as shown in Figure 3.3.

Backing up data is another routine task that should be carried out periodically to ensure that data is secure. There are a number of ways to back up data, both online and offline, and a range of media that can be used to back up data.

Activity 3.5

When you work with data, you should always ensure that you have saved your files and back up appropriately.

Try to get into the routine of applying good practice by ticking off items on the checklist to remind you to:

Procedure	Criterion	Adhered to	
		Yes ✓	No ✗
File management	Save work regularly		
	Use sensible filenames		
	Choose appropriate file formats		
	Set up directory/folder structures to organize files		
	Used a suitable backup procedure		

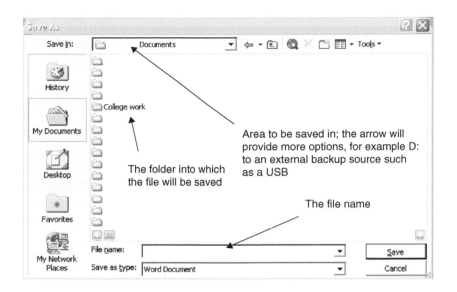

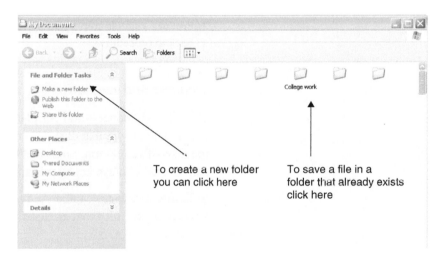

Figure 3.3 Naming files and using folders.

Backup strategies will protect your data from a range of mishaps, including:

- accidental changes to or deletion of data
- loss of data due to media or software faults
- virus infections
- hackers
- catastrophic events (fire, flood, etc.).

Therefore any strategy should take into consideration the following:

- frequency
- multiple copies
- offsite backup
- media.

To ensure that backups remain up to date the backup procedure should be carried out regularly. The greater the frequency of data being changed the more often backups should be made. If data is changing significantly every day this should be reflected in the frequency, i.e. daily backups.

CHAPTER 3

As well as backing up frequently, several backup copies should be made at different dates in case an undiscovered problem from a working copy arises.

Some backup copies should be stored 'offsite' to safeguard against more serious threats such as fire or physical disasters.

Backup copies should be made on new media. Problems may occur if media are used more than once, and faults may start to develop, for example floppy disks are not a good medium for backup copies. If they are used, they should be replaced frequently. Backup copies should also be stored on multiple media (e.g. zip disk and CD-ROM) to avoid all backup copies becoming corrupted by the same drive or disk fault.

Activity 3.6

1. Demonstrate that you can carry out a range of file management and backup procedures.

2. Produce a simple step-by-step guide on saving data, creating folders and backup procedures.

Cleaning and ventilation

Cleaning components and peripheral items such as printers and mice is essential to ensure that dust does not build up in areas that could restrict the flow of ventilation. Specialist tools and miniature vacuum cleaners can be used to clean keyboards and other hardware items.

Solutions and wipes can be used to ensure that items such as monitors, keyboards and mice that are used frequently are kept clean and free from bacteria.

Maintaining systems

There are a number of routine tasks that can be carried out on computer systems. In terms of maintenance these tasks can consist of:

- Replacing consumables such as paper, ink or toner cartridges
- Replacing damaged components.

One of the key considerations for replacing consumables and components is the correct disposal of these items. The drive towards recycling cartridges and safe disposal of components has led to an increase in organizations that offer recycling provisions and incentives to individuals and organizations that embark upon this process.

Be able to monitor systems and perform necessary enhancements to systems for improvements in performance and efficiency

Monitoring

Several diagnostic tools and utilities are available to undertake monitoring tasks on a system. One example is Registry Mechanic. This provides specific repair and optimization tools for the Windows® registry. Diagnostic tools and utilities can be used to support a range of hardware and software components. Some will perform isolated tasks to detect faults, clean up files and folders, or improve the efficiency, speed and productivity of a system.

Server management

The simple network management protocol (SNMP) is used in network management systems to monitor any attached devices on the network for conditions that may require administrative attention.

Remote administration

Some organizations opt for remote support and administration. This service is driven by software that allows remote administration. It can be beneficial when it is impractical to carry out administrative duties physically near to the system.

Improving system performance

ROM-BIOS is an abbreviation for read only memory–basic input output system. In PCs the BIOS contains all the code required to control the keyboard, monitor/display, disk drives, serial port(s), and a number of other low-level functions, e.g. memory timing.

The BIOS is typically placed in a ROM chip that comes with the computer. This ensures that the BIOS is always available and cannot be damaged by disk failures. The PC BIOS is fairly standardized, so all PCs are similar at this level (although there are different BIOS versions). Additional DOS functions are usually added through software modules. This means you can upgrade to a newer version of DOS without changing the BIOS.

A computer's BIOS holds details of the hardware that is installed and the settings for individual components. When the computer starts up, the BIOS starts the power-on self-test (POST) routine. This routine checks the motherboard and its components to make sure that they are operating normally, and the hardware that is installed is checked to see that it matches the system's BIOS. If an error is identified during this startup process the system reports it to the user via an error message or 'beep'.

Parameters that are defined within the ROM-BIOS include being able to select startup (boot) disk drive, setting a system password, defining a new disk drive and configuring a new card, e.g. a video card.

The BIOS setup programme should also be run after power-on to configure specific settings for any new hardware that has been installed.

During the POST routine there is usually a key that needs to be pressed to start the BIOS setup programme. Depending on the make of BIOS this will normally be one of the following keys: F1, F10, Escape, Delete or Insert. Sometimes a message appears on screen to tell you which key to press.

Once the setup programme has started specific settings for any new hardware can be entered.

CHAPTER 3

ICT systems can be configured to start up and operate in different ways. Much of this can be controlled by the ICT system manager; however, some can be configured to suit the needs of the end-user. You will be expected to set up different systems to meet the needs of an end-user. You will also be expected to use the operating system for settings and diagnostics such as:

- time and date
- printer, mouse and keyboard configuration
- password properties
- multimedia configurations
- scheduled tasks
- graphical user interface (GUI) desktop and display setup
- virus protection configuration
- directory (folder) structure and settings
- applications software icons
- checking and setting system properties
- power management.

The majority of configurations can be carried out through the control panel on the operating system, although some, such as the setting of the time and date, can also be done through the ROM-BIOS startup.

In the image featured on the control panel you can see how the majority of settings can be adjusted to suit user needs (Figure 3.4a–d). These include the time and date, printer, mouse and keyboard configurations,

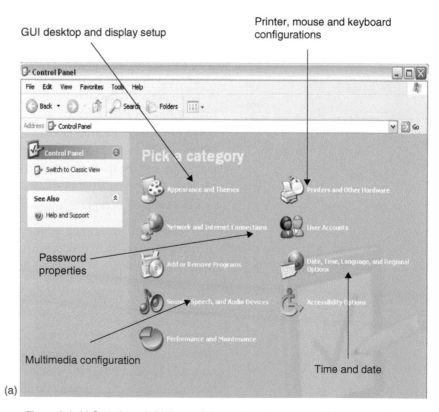

(a)

Figure 3.4 (a) Control panel; (b) time and date settings; (c) printer configurations; (d) password properties.

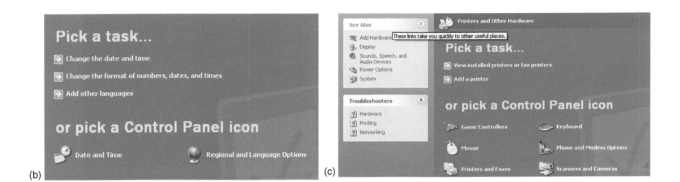

(b)   (c)

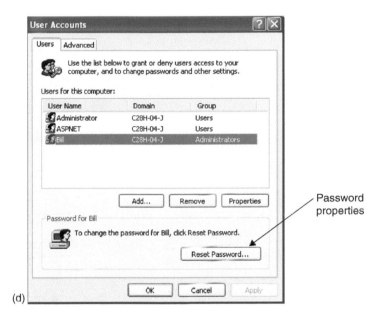

(d)

Figure 3.4   (Continued).

and password properties. User accounts can also be set up to protect and restrict logins to certain applications.

Multimedia configurations such as adjusting the graphics and sound can be addressed through the control panel.

Scheduled tasks such as backing up data can be carried out through the operating system. Depending on whether you are using a networked or standalone computer, the backup procedure will download your data onto a local hard disk or onto the network, where specific storage areas are designated for individual users. With most graphical operating systems procedures for carrying out scheduled tasks involve either 'point and click' or 'drag and drop'. On the performance and maintenance menu in the control panel (Figure 3.5a) you can also carry out a range of scheduled tasks.

Changing the screen resolution of the monitor is another example of checking and setting system properties (Figure 3.5b). Power management functions that control your ability to log off, switch between users and shut down can be accessed through the 'Start' menu (Figure 3.5c).

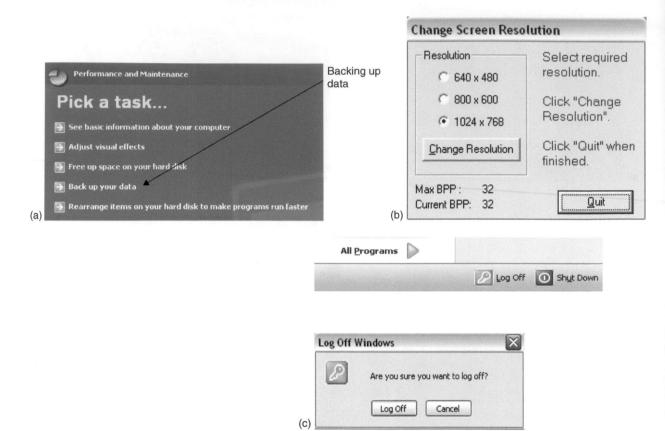

Figure 3.5 (a) Backing up data; (b) changing the screen resolution; (c) logging off.

Firmware updates

Firmware sits somewhere between hardware and software. It is a computer programme that is embedded within a hardware device; however, it can also be provided on flash memory or as a binary image file.

Examples of firmware include:

- the BIOS
- platform code
- certain control menus, e.g. those found on iPods
- open firmware used in Sun Microsystems and Apple Macs.

Operating system settings

A system's performance can be improved by adjusting the settings on the operating system, in conjunction with the control panel.

Through the operating system you can change the way the GUI desktop and display setup looks. This can be done through the 'Appearance and themes' setting on the control panel.

Virus protection software can be installed or downloaded from the Internet and configured to start up every time you boot up your computer. Reminders will also be activated to enable you to check and update the version and level of virus protection at regular intervals.

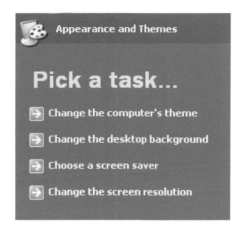

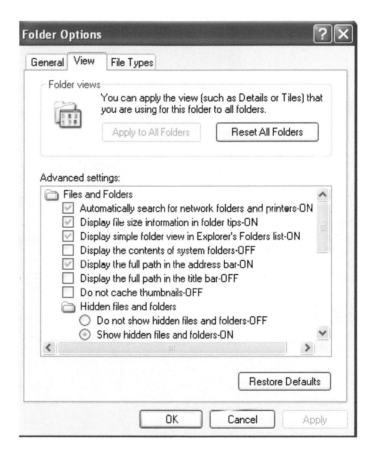

Figure 3.6 Directory (folder) structure and settings.

Access to directories and folders can be gained from the 'Start' menu in operating systems such as Windows, where you can look into 'My computer' or 'My documents' to locate directories and folders. 'Folder options' can also be viewed through the control panel (Figure 3.6).

Applications software icons can be changed through the 'properties' function, by right-clicking on the icon. In addition, icons can be downloaded and used to replace those that are supplied with the operating system (Figure 3.7).

CHAPTER 3

Figure 3.7 Application software: traditional icons.

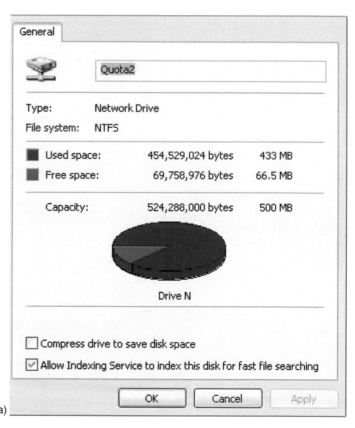

(a)

Figure 3.8 Disk optimization.

Disk optimization

Viewing information about the hard drive, such as checking how much space is available on the hard disk, through 'properties' (Figure 3.8), and optimizing resources to free up space is one way of improving system

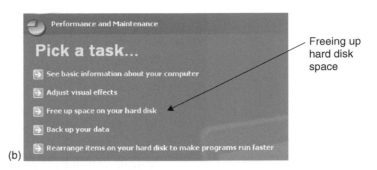

Freeing up
hard disk
space

(b)

Figure 3.8 (Continued).

performance. Defragmenting and running disk error checks will also improve the overall performance.

System protection

All systems should be protected by antivirus or antispyware software. System protection can improve efficiency and enhance overall performance.

'System restore' can be used to return the system to its previous state, before making a configuration change. This function is integrated within Vista and XP and offers protection in the form of a failsafe mechanism.

Upgrading

Upgrading components such as processors, memory, video cards and motherboards can improve the performance of a system, especially if the components to be replaced are ineffective or redundant and require a faster or more efficient model or version.

Software upgrades may include the reinstallation or application of patches to fix known bugs or faults. Installation procedures could fix errors that have developed and may resolve issues such as system crashes.

System rollback

System rollback through the use of software is one way to restore your system and aids any disaster recovery procedure. Software can be used to restore the system back to its 'clean' state. Vista and XP offer a 'system restore' tool that returns the system to its previous state before making a configuration change.

Drawbacks and benefits

There are several drawbacks and benefits to upgrading hardware and software. If this is a non-scheduled activity it could impact on users by causing delays to their work schedules. If hardware or software needs to be purchased which is not part of the procurement process additional money and resources may have to be deployed.

Documentation and review

A range of documentation can be used to support system upgrades. Documentation will be required to track and audit systems that have been identified as requiring an upgrade. Purchase order documents or tender invitations may be issued if the upgrade is on a large scale. Schedules for the upgrade process will be required that identify times and tasks and the impact that the process may have on end-users. Service-level agreements may be drawn up if the upgrades are being conducted by a third party provider. Finally, once the upgrade has been completed evaluation and review documents may need to be produced and signed off.

Testing functionality

Testing procedures should provide the framework for any upgrade or installation. Testing will need to be carried out in stages to check that individual components installed or software upgrades are functioning properly and that any fault diagnosis previously recorded has been addressed and eliminated.

Questions and review

1. Why is it important to plan and perform computer system maintenance?
2. Provide examples of routine and non-routine problems.
3. What is meant by the term 'procurement'?
4. Under what circumstances might an escalation procedure need to be enforced?
5. Provide examples of two types of planning tools that can be used to support the computer system maintenance procedure.
6. Identify at least three pieces of legislation that are appropriate and should be adhered to when undertaking computer maintenance-related tasks.
7. What measures can be taken to reduce electrostatic charge (ESD)?
8. Identify a range of routine housekeeping tasks that can be carried out on computer systems.
9. What types of monitoring tools and utilities can be used to enhance system efficiency or performance?
10. What are the benefits and drawbacks of upgrading the hardware within a system?

Assessment activities

Grading criteria	Content	Suggested activity
Pass		
P1	Explain the need to plan scheduled routine and non-routine computer systems maintenance.	Produce an information pack for end users that could incorporate a number of leaflets. The first leaflet could include information for P1 that explains the need to plan scheduled routine and non-routine computer systems maintenance.
P2	Describe health and safety risks facing the user and practitioner whilst working with computer systems, identifying for each one appropriate legislative guidelines and recommended precautions.	A health and safety booklet/leaflet could also be produced to go into the pack that incorporates P2. Describe health and safety risks facing the user and practitioner whilst working with computer systems. In addition to this appropriate legislative guidelines and recommended precautions should also be identified.
P3	Identify housekeeping procedures that need to be performed on computer systems.	Draw up a checklist of housekeeping procedures that need to be performed on computer systems.
P4	Identify an upgrade opportunity for hardware and one for software through use of monitoring tools.	Identify an upgrade opportunity for hardware and one for software through use of monitoring tools. This can be carried out in conjunction with P6.
P5	Perform routine housekeeping on a computer system.	Demonstrate that you can perform routine housekeeping on a computer system. This demonstration can also be supported by documentary evidence such as a checklist, witness statement or observation sheet.
P6	Upgrade hardware and check functionality.	In conjunction with P5, demonstrate that you know how to upgrade hardware and check functionality.
P7	Comply with safe working practices whilst maintaining computer systems.	You should demonstrate that you are complying with safe working practices whilst maintaining computer systems. The evidence for this could include adhering to a number of safe working practices and confirming this through a practical demonstration or observation.
Merit		
M1	Explain the need for policies and procedures to control the maintenance of computer systems activities in organizations.	As part of the information pack for P1 and P2, include policy and procedure information that could be used to control the maintenance of computer systems activities in organisations.
M2	Recommend one possible hardware upgrade and one possible software upgrade based on their respective benefits and drawbacks.	In conjunction with P6 you could also provide a written proposal aimed at a particular client or user that recommends one possible hardware upgrade and one possible software upgrade based on their respective benefits and drawbacks.
M3	Explain the sustainability and environmental issues that relate to the maintenance and upgrading of computer systems.	As part of the information pack you could also include information about sustainability and environmental issues that relate to the maintenance and upgrading of computer systems. Provide a case study or evidence of research to support this evidence.
Distinction		
D1	Discuss and evaluate improvements to computer systems achieved by routine housekeeping procedures.	In conjunction with P3, provide a written evaluation that discusses the improvements that can be made by using routine housekeeping procedures as identified in the checklist for P3.
D2	Evaluate performance changes to computer systems through performing and documenting chosen hardware and software upgrades.	In conjunction with any of the practical activities, evaluate performance changes to computer systems through performing and documenting chosen hardware and software upgrades.

CHAPTER 3

Courtesy of iStockphoto, kohlerphoto, Image#2676290

Networks have enhanced the way in which organisations can communicate, improving data and information exchange between internal and external users, employees and stakeholders. Managing these networks extends from general maintenance right through to ensuring that the network is secure.

Chapter 4

Network Management

The management of networks is essential to ensure that they function fully and that they are kept secure from a range of internal and external threats. Network management can be left to the responsibility of a single person, a team of IT support personnel or specialist software that can monitor, filter and support users in a range of network tasks.

This chapter will provide you with an overview of networked systems and the emerging technologies that can assist and impact on networks and the maintenance, configuration and customization of these systems.

The chapter is structured around the following learning outcomes:

- Understand the principles of network management.
- Know about networking management tools and technologies.
- Be able to carry out network management activities.

Understand the principles of network management

Network management is extremely important to the security, operation and maintenance of any network. Network management can cover a range of areas and tasks that include the physical hardware, software, end-users, technical, security and data elements of the system, as explored further within this first section.

Network management functions

Networks are designed to meet a range of user and organizational functions, that range from sharing and utilizing system resources, through improving response times and performance, to ensuring that data is kept secure and information and processes are monitored and auditable.

Network management functions can include a range of the following activities:

- configuration
- fault management
- management/account management
- performance variables, such as network throughput, user response times and line utilization
- other activities, including planning, designing and installing
- network operations, for example security, data logging, checking performance and traffic
- reporting.

Configuration and management configuration deal with a multitude of tasks associated with adapting, updating, controlling and fixing hardware, software and documentation to ensure that networked systems are robust and working productively and efficiently.

Fault management functions can identify, isolate and even rectify any errors or malfunctions that occur within a system. Fault management can be performed at two levels: actively and passively. Active fault management uses monitoring devices such as 'ping' to determine whether a device is active and responding. Passive fault management works by gathering alarms from devices when something happens within the devices.

The management of networked systems ensures that everything is monitored to ensure peak performance, maintain reliability, stability and efficiency of data, systems and users.

Performance variables are used to monitor the quality of service and utilization of a network in terms of when and how it is used, peak times of usage, user response times and line utilization.

Network functions can also include planning, designing and installing of hardware, software, users, systems, projects and other activities.

Network security, data logging and the checking of performance and traffic are vital operations that are required to ensure that a

CHAPTER 4

network operates effectively and securely. Monitoring and auditing of data, users and systems is vital to maintain data protection across a network.

Reporting is an important element of any system. Data can be gathered about system performance, utilization, service provision, usage and other criteria to identify patterns or trends or to forecast and make value judgements about what will be required now and in the future.

Network operating systems

A number of network operating systems is available on the market, examples of which include Windows Server 2008, 2003 or 2000, Netware, Linux, Unix and IBM i5/OS.

Network operating systems, like normal operating systems, are essential for network startup and operation. The type of network operating system selected will vary depending on the user or size of an organization, i.e. scalability. For example, a small organization running a peer-to-peer network might use Windows networking such as XP. A large organization, however, may have more complex networking requirements that would warrant the use of a more scalable network operating system such as Windows 2008, 2003 or 2000. Alternatively, they may be operating a combination of different network operating systems according to the platform and performance needs.

Activity 4.1

A number of network operating systems is available.

1. Analyse at least one network operating system and produce a table that examines a range of its tools and functions. Comment on its functionality both for an organization and from an end-user's perspective.

Network operating system: ....

Version: ....

Functions (what can it do?)	Tools and features	How does it support network management features?	How does it support end-users?	Other

---

Networking protocols

Communication between different devices requires agreement on the format of the data. The set of rules that define the format is known as a 'protocol'.

Protocols can be incorporated in either the hardware or the software. They are arranged in a layered format (sometimes referred to as a protocol stack) as shown in Figure 4.1. Protocols can provide some or all of the services specified by a layer in the open systems interconnection (OSI) model.

Middleware – a piece of interface software between two or more different systems that allows the systems to communicate.

CHAPTER 4

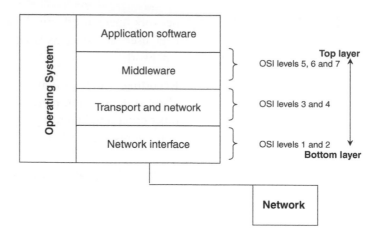

Figure 4.1 Protocol layers.

The OSI layer model

When setting up a network correctly you need to be aware of the major standards organizations and how their work can affect network communications. In 1984 the International Organization for Standardization (ISO) released the OSI reference model, which has subsequently become an international standard and serves as a guide for networking procedures and visualizing networking environments.

The model provides a description of how network hardware and software can work together in a layered framework to promote communications. The model also provides a frame of how components are supposed to function, which assists with troubleshooting problems.

The OSI reference model divides network communication into seven layers. Each layer covers different network activities, equipment or protocols as shown in Table 4.1.

Table 4.1 OSI reference model

Level	Description
7	Applications layer
6	Presentation layer
5	Session layer
4	Transport layer
3	Network layer
2	Data link layer
1	Physical layer

- **Physical layer** – provides the interface between the medium and the device. The layer transmits bits and defines how the data is transmitted over the network. It also defines what control signals are used and the physical network properties such as cable size and connector.
- **Data link layer** – provides functional, procedural and error detection and correction facilities between network entities.
- **Network layer** – provides packing routing facilities across a network.

- **Transport layer** – an intermediate layer that higher layers use to communicate to the network layer.
- **Session layer** – the interface between a user and the network, this layer keeps communication flowing.
- **Presentation layer** – ensures that the same language is being spoken by computers.
- **Applications layer** – ensures that the programmes being accessed directly by a user can communicate, e.g. an e-mail programme.

Transmission control protocol/Internet protocol

Transmission control protocol/Internet protocol (TCP/IP) is the standard protocol used for communication among different systems. TCP/IP also supports routing and is commonly used as an Internet working protocol.

Other protocols written specifically for the TCP/IP suite include:

- f le transfer protocol (FTP) – for exchanging files among computers that use TCP/IP
- simple mail transfer protocol (SMTP) – for e-mail
- simple network management protocol (SNMP) – for network management.

There are many advantages of TCP/IP, including:

- **Expandability** – because it uses scalable cross-platform client/server architecture it can expand to accommodate future needs.
- **Industry standard** – being an open protocol means that it is not managed or controlled by a single company, therefore it is less prone to compatibility issues. It is the de facto protocol of the Internet.
- **Versatility** – it contains a set of utilities for connecting different operating systems, therefore connectivity is not dependent on the network operating system used on either computer.

Activity 4.2

1. What is meant by a 'protocol'? Why are protocols so important in networking?
2. What do the following stand for:

   - OSI
   - SMTP
   - TCP/IP
   - SNMP?
3. How many layers are there in the OSI model? What are they?
4. Give one advantage of TCP/IP.

Design considerations

Security

Keeping data secure can be difficult because of the environment in which users work and levels of user and access requirements to the data. With the movement towards a totally networked environment promoting

CHAPTER 4

a culture of sharing, the issue of data security is even more important and should be addressed at a number of levels (Figure 4.2).

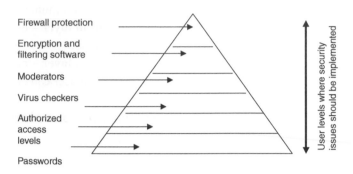

Figure 4.2 Levels of security.

As illustrated in the figure, security measures need to be integrated at each user level within an organization. The indication of security measure does not confine it to a certain level but reflects on an organizational scale what should be implemented and the scale of implementation. In addition to these proposed security measures there is the issue of physical security – ensuring that hardware and software are kept physically secure under lock and key.

The actual protection of data can be resolved quite easily by introducing good practice measures such as backing up all data to a secondary storage device, limiting file access, and imposing restrictions to read only, execute only or read/write. However, data protection is also covered more widely under certain Acts such as the Data Protection Act 1998.

Firewall protection

The primary aim of a firewall is to guard against unauthorized access to an internal network. In effect, a firewall is a gateway with a lock, which only opens for information packets that pass one or more security inspections.

There are several different firewall types:

- **Application gateways** – the first gateways, sometimes referred to as proxy gateways. These are made up of hosts that run special software to act as a proxy server. Clients behind the firewall must know how to use the proxy, and be configured to do so to use Internet services. This software runs at the application layer of the ISO/OSI reference model, hence the name. Traditionally, application gateways have been the most secure, because they do not allow anything to pass by default, but need to have the programmes written and turned on to begin passing traffic.
- **Packet filtering** – a technique whereby routers have 'access control lists' turned on. By default, a router will pass all traffic sent to it, and will do so without any sort of restrictions. Access control is performed at a lower ISO/OSI layer. Since packet filtering is done with routers, it is often much faster than application gateways.

- **Hybrid system** – a mixture of application gateways and packet filtering. In some of these systems, new connections must be authenticated and approved at the application layer. Once this has been done, the remainder of the connection is passed down to the session layer, where packet filters watch the connection to ensure that only packets that are part of an ongoing (already authenticated and approved) conversation are being passed.
- **Stateful firewall** – keeps track of the state of network connections travelling across it. Includes any firewall that performs stateful packet inspection (SPI) or stateful inspection.

Encryption and filtering software

Encryption software scrambles message transmissions. When a message is encrypted a secret numerical code, the 'encryption key', is applied, and the message can be transmitted or stored in indecipherable characters. The message can only be read after it has been reconstructed through the use of a 'matching key'.

Moderators

Moderators have the responsibility of controlling, filtering and restricting information that is shared across a network.

Virus checkers

These programmes are designed to search for viruses, notify users of their existence and remove them from infected files or disks.

Authorized access levels and passwords

On a networked system various privilege levels can be set up to restrict users' access to shared resources such as files, folders, printers and other peripheral devices. A password system can also be implemented to divide levels of entry in accordance with job role and information requirements.

For example a finance assistant may need access to personnel data when generating the monthly payroll. Data about employees, however, may be password protected by personnel in the human resources department, so special permission may be required to gain entry to this data.

Audit control software allows an organization to monitor and record what they have on their network at a point in time and provide them with an opportunity to check that what they have on their system has been authorized and is legal.

Over a period of time a number of factors could impact on how much software an organization acquires without their knowledge. These can include:

- illegal copying of software by employees
- downloading of software by employees
- installation of software by employees
- exceeded licence use of software.

CHAPTER 4

These interventions by employees may occur with little or no consideration to the organization and its responsibility to ensure that software is not being misused or abused.

User rights and file permissions

Within certain IT systems, users are given permissions to access some areas of a folder, application or document and are restricted from others. By allowing users certain rights within a given system, security of data can be reassured and the span of control can be limited. An example of this can be seen in the case of an IT system in a doctor's surgery. The administration staff may have access to appointments and scheduling, the nurses may have access to patient information, and the GP could have full access and rights to print out prescriptions and authorize medication.

Certain permissions may also be set up to allow certain users partial access to a file, so for example information can be read (read only), but not written to, or users may be able to run a programme (execute only) but not view it. Users with full read/write permissions would be able to view, update, amend and delete accordingly.

Know about networking management tools and technologies

Networking management tools and technologies can be used to support organizations, network managers, administrators, technical personnel and end-users in their day-to-day networking tasks. Networking management tools can be used to identify and diagnose faults, monitor performance and service levels, and ensure that the network is operating in an efficient and secure environment.

Network devices

Various devices are required to ensure the smooth functioning and operation of a network. These devices can be categorized as:

- servers
- workstations
- interconnection devices
- network cards
- vendor-specific hardware.

Servers

The server is a powerful computer that stores the application and the data that is shared by users. Servers effectively circulate the information around the network and together with the network operating system perform a number of functions (Figure 4.3).

Applications and data can be managed more effectively when they are managed by a server. Auditing functions can also be undertaken more easily to ensure that data is being kept secure.

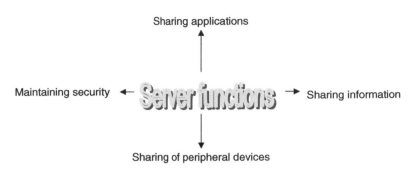

Figure 4.3 Server functions.

All of the machines on the Internet are either servers or clients. The machines that provide services to other machines are servers, and the machines that are used to connect to those services are clients. Servers can be categorized into the following:

- web servers
- print servers
- e-mail servers
- FTP servers
- newsgroup servers
- storage servers.

When you connect to a website to read a page, you are accessing that site's web server.

Workstations

A workstation is a high-end desktop or microcomputer that is designed to provide and deliver a technical service provision, as illustrated in Case Study 4.1. Workstations are more powerful than everyday personal computers as they are designed to carry out more complex and multitasking activities.

Interconnection devices

Switch

A switch can connect Ethernet or other types of packet switched network segments together to form a diverse network that operates within the data link layer of the OSI layer model and sometimes the network layer. A switch can also manage traffic at layers 2, 3 and 4.

Routers

Routers determine where to send the information from one computer to another. They are specialized computers that send the messages quickly to their destination(s) along thousands of pathways.

A router serves two purposes; first, it ensures that information does not go astray, which is crucial for keeping large volumes of data from clogging up connections, and secondly, it ensures that the information makes it to the intended destination.

CHAPTER 4

Case Study 4.1

HP Blade Workstation Solution

Secure, high performance and flexible workstation computing

Solution is the only solution of its kind: a next generation workstation infrastructure that combines the centralized control and security of the data center with a graphics intensive, workstation-class experience and the flexibility to support remote professionals. It is an innovative offering built with the HP ProLiant quality and trading floor proven reliability that demanding customers like you expect from HP.

The HP Blade Workstation Solution is a complete product offering that enables you to centralize the power of your workstation environment for remote user access and collaboration without experiencing any loss in performance.

It's time to revolutionize your workstation environment.

Data center security and control

The HP centralized blade workstation approach brings mission-critical security & operational continuity to workstation computing. High-performance blade workstations send encrypted graphic data to client devices. This minimizes the risk of security exposures from local hard drives, removable media drives and data interfaces like USB, as well as through system theft or loss. Multi-blade, multi-site capabilities also improve business continuity and the management toolset gives your IT staff efficient, expert control.

Cool and quiet Workstation power

The new HP Blade Workstation Solution enables you to access workstation compute and graphics power from a blade client device on-demand. This device is optimized for security, low noise and low heat emission, which radically changes your work experience. This solution makes it possible to remotely share and deliver advanced workstation graphics, in 2D or 3D, and motion-picture quality and full motion video. The solution can leverage your existing Windows-based, networked devices, such as thin clients, PCs and notebooks, and allows you to use your existing applications.

http://h41112.www4.hp.com/blades/uk/en/

In performing these two roles a router is invaluable. It joins networks together, passing information from one to another, and also protects the networks from each other. It prevents the traffic from one network unnecessarily spilling over to another.

Regardless of how many networks are attached, the basic operation and function of the router remains the same. Since the Internet is one huge network made up of tens of thousands of smaller networks the use of routers is an absolute necessity.

In order to handle all the users of even a large private network, millions and millions of traffic packets must be sent at the same time. Some of the largest routers are made by Cisco Systems Inc. and Juniper.

Wireless

Wireless communication facilitates the transfer of information between fixed or mobile devices without the use of wires or electrical conductors. Wireless communication may be through the use of radio frequency communication, microwave or infrared.

Bluetooth

Bluetooth facilitates the exchange of data over short distances between mobile devices such as mobile phones, personal digital assistants (PDAs) and netbooks, or fixed devices, using a wireless protocol.

Network cards

A network card, sometimes referred to as a network interface card (NIC), plays a very important role in connecting a cable modem and a computer together. It enables the user to interface with a network through either wires or wireless technologies. The network card allows data to be transferred from one computer to another computer or device.

Vendor-specific hardware

Vendor-specific hardware can include components and systems from organizations such as Cisco, HP and 3Com (Figure 4.4). Organizations or vendors often supply the necessary hardware to ensure compatibility among their own systems, however problems can occur when using

Figure 4.4 Vendor-specific hardware providers.
Screenshots taken from the Cisco, HP and 3COM websites.
http://www.cisco.com, http://www.hp.com, http://www.3com.com/

hardware and software mixes from different vendors, which may force an organization down a single vendor route.

Networking tools

Networking tools are designed to support the network managers and administrators in their roles of supplying a top-end quality networking provision to a range of stakeholders and end-users. Networking tools can provide support in the areas of fault management and performance management, using tools such as HP Openview and Cisco Works.

Activity 4.3

1. Research information about HP Openview and Cisco Works and produce a one-page information sheet about the two.
2. Try to identify an organization that uses either Openview or Works and discuss how this networking tool provides support

Layout

The layout of a network is based around the planning and framework of the size, extent, capacity and overall network functionality. Most of these considerations can be addressed through hardware and software, however consideration also needs to be given to the cabling type and standards, as well as the network topology.

Standards in cabling are based on three types:

- **Cabling EIA/TIA 568A** – this is the American standard and was the first to be published in 1991.
- **ISO/IEC 11801** – the international standard for structured cabling systems.
- **CENELEC EN 50173** – the European cabling standard (the British version is BS EN 50173).

Standards are required to define a method of connecting all types of vendors' voice and data equipment over a cabling system that may use common connectors media or a common topology.

The main types of cables used in networking (Table 4.2) are:

- unshielded twisted pair (UTP) (Figure 4.5)
- shielded twisted pair (STP)
- coaxial
- f breoptic
- Cat5e, Cat 6 and Cat 7.

Table 4.2 Cat cabling comparison

Specification	Cat5e	Cat6	Cat7
Frequency (MHz)	100	250	600
Attenuation (dB)	24	19.8	20.8

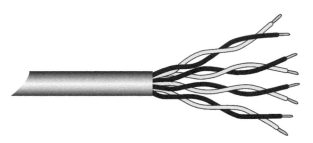

Figure 4.5 Unshielded twisted pair (UTP) cable.

The standards for UTP cabling extend from category 1 to 5, where category 1 is voice only and category 5 provides a fast Ethernet provision.

Shielded twisted pairs are often used on token ring topology networks.

Coaxial cables are often used as a transmission line for radio frequency signals. The cable consists of an inner conductor, surrounded by an insulating layer, a conductive layer and finally a thinner insulating layer on the outside.

Fibreoptic cabling (Figure 4.6) has the ability to transmit signals over longer distances than twisted pair or coaxial cables. The fibreoptic technology transmits light rather than electronic signals.

Figure 4.6 Fibreoptic cabling.

Cat5e cabling is often used in computer networks such as Ethernet and can also be used to carry basic voice services, token ring and asynchronous transfer mode (ATM). It can provide transmission speeds of up to 1000 Mbps. Cat6 cabling is similar to Cat5e but it offers a higher standard, and similarly Cat7 cabling offers a higher standard than Cat6.

Attenuation – any reduction in signal strength.

Activity 4.4

1. Students should be introduced to a range of different network cables, including:
    - shielded twisted pair (STP)
    - unshielded twisted pair (UTP)
    - thick Ethernet (thicknet)
    - thin Ethernet (thinnet)
    - fibreoptic cables.

2. Based on these different cable types, students should produce a table identifying what each cable is used for and the properties and characteristics of each cable type.

Activity 4.5

(Tutor-led practical activity)

1. Discuss and demonstrate how various cables can be assembled.
2. Explain that this session has been set up to give students a more practical insight into network cabling.
3. Discuss any protocols involved with cable making and ensure that students are aware that health and safety procedures must be adhered to during the practical.
4. Some students will be in a position where they may not have assembled a cable before. Engage students in this process by setting up the resources to demonstrate how this can be achieved.
5. Set up an environment and demonstrate how to put a cable together. You may have to bring small groups of students up at a time to ensure that they can see clearly what steps are being followed.
6. Provide students with an opportunity to make their own cables and test them. Ensure that students take notes of the stages/steps involved.
7. The session should end with a review and consolidation, followed by any questions about the activities undertaken.

Topologies

Networks vary in size and complexity. Some are used in a single department or office, while others extend across local, national or international branches. Networks vary in structure, to accommodate the need to exchange information across short or wide geographical areas. These structures include:

- local area networks (LANs)
- metropolitan area networks (MANs)
- wide area networks (WANs) – long-haulage networks (LHNs)
- value-added networks (VANs)
- personal area networks (PANs).

Local area networks

These consist of computers that are located physically close to each other, within the same department or branch. A typical structure includes a set of computers and peripherals linked as individual nodes. Each node, for example a computer and shared peripheral, is directly connected by cables that serve as a pathway for transferring data between machines.

Metropolitan area networks

These are more efficient than a LAN and use fibreoptic cables to allow more information and a higher complexity of information. The range of a MAN is also greater than a LAN, allowing business to expand; however, this can prove to be expensive because of the fibreoptic cabling.

Wide area networks – long-haulage networks

These are networks that extend over a larger geographical distance from city to city within the same country or across countries and even continents. WANs transfer data between LANs on a backbone system using digital, satellite or microwave technology. A WAN connects different servers at a site. When this connection is from a personal computer (PC) on one site to a server on another it is referred to as being 'remote'. If this coverage is international it is referred to as being an 'enterprise-wide network'.

Value-added network

This type of network is a data network that has all the benefits of a WAN but with vastly reduced costs. The cost of setting up and maintaining this type of network is reduced because the service provider rents out the network to different companies, rather than an organization having sole ownership or a 'point-to-point' private line.

Networks can be used to support a range of applications within an organization. The selection of a particular network depends on:

- the application/use
- the number of users requiring access
- physical resources
- the scope of the network – within a room or department, across departments or branches.

If, for example, a network was required to link a few computers within the same department to enable the sharing of certain resources, a LAN might be installed.

If, however, a network was required to link branches and supplier sources across the country, a WAN might be installed, with the server located at the head office providing remote access to users connecting at individual branches.

Personal area network

A PAN is a short-range network that is used for communication among computer devices, with a typical range of a few metres. PANs can be wired or wireless (WPAN).

Activity 4.6

1. What do LAN, MAN and WAN mean?
2. What are the differences between a LAN, a WAN and a MAN?
3. Do you think a VAN is better than a WAN?
4. Provide an example of a PAN.
5. What type of network might you consider if you wanted to link four computers together in a single office?

Topology refers to the layout of connected devices on a network. Network topologies include:

- bus
- ring
- star
- tree
- mesh.

CHAPTER 4

A bus topology is based on a single cable which forms the backbone to the structure with devices that are attached to the cable through an interface connector (Figure 4.7). When a device needs to communicate with another device on the network a message is broadcast. Although only the intended receiving device can accept and process the message, a disadvantage of this topology is the fact that other devices on the network can also see the message.

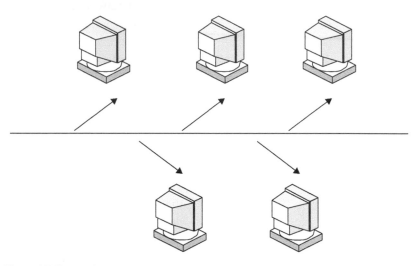

Figure 4.7 Bus topology.

In a ring topology every device has two neighbours for communication purposes (Figure 4.8). Communication travels in one direction only, either clockwise or anticlockwise. One disadvantage of this type of topology is that a failure in the cable can disable the entire network.

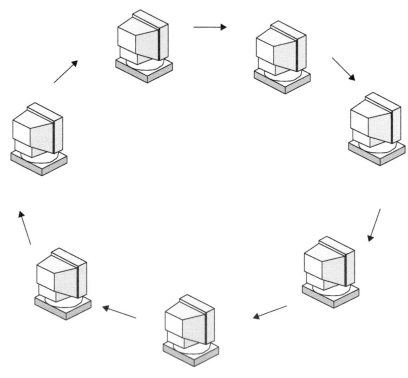

Figure 4.8 Ring topology.

Star topologies are based around a central connection point, referred to as a hub or a switch (Figure 4.9). A series of cables is used in this type of network, with the advantage that if one cable fails only one computer will go down and the remainder of the network will remain active. If the hub fails, however, the entire network will fail.

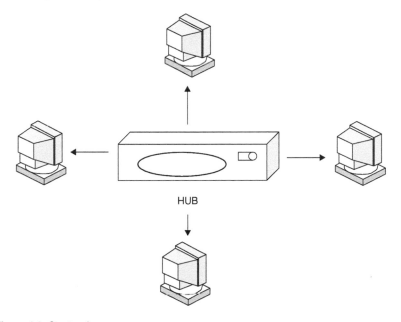

Figure 4.9 Star topology.

Tree topologies integrate a number of star topologies together onto a bus, with the hub devices connecting directly to the tree bus, each hub acting as the 'root' of the tree device.

A mesh topology introduces the concept of 'routes', where several messages can be sent on the network via several possible pathways from source to destination.

Client/server systems

The client/server approach is based around a client, which is responsible for processing requests for data, and a server, which executes the request and returns its results (Figure 4.10).

The client element requires 'intelligence' (e.g. memory), therefore most clients are PCs. A 'dumb' terminal would not work because it lacks memory. The PCs are access points for end-user applications.

Emerging technologies and their impact

Emerging technologies in any field can have a tremendous impact on the way in which systems are designed and function. Emerging technologies can also affect communications and the way in which users interact with hardware, software, data and information.

Emerging technologies can include technologies such as mobile networking, web interfacing and remote monitoring, such as remote network monitoring (RMON). These technologies bring a whole package of benefits to the organization, end-user and customer (Figure 4.11).

CHAPTER 4

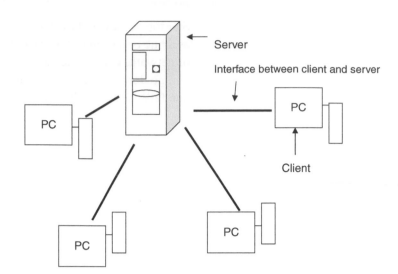

Figure 4.10 Example of a client/server.

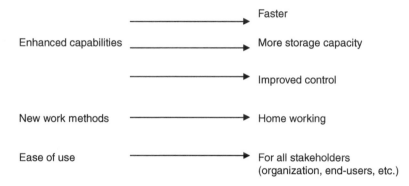

Figure 4.11 Benefits and impact of emerging technologies.

Be able to carry out network management activities

Network management activities can include a wide range of tasks, processes and functions. A number of practical activities can be carried out to prepare you for this task.

Throughout this chapter information has been provided on network functions, network management tools and technologies. All of this information can be used to provide the knowledge and skills needed to carry out network management activities.

The following sections will provide you with opportunities to attempt practical activities that will prepare you for your unit assessments.

Maintenance activities

Students should be introduced to a range of regular maintenance activities, including:

- how to back up and restore files
- creation and deletion of user accounts
- designing and developing log-on scripts

- conducting virus scans
- performing file cleanups
- using tools to manage performance and fault finding.

Backing up files means copying files or file systems to removable media to safeguard against loss, damage or corruption. Restoring files means copying current backup files from removable media to a working directory.

Users who log in and interact with a network will require their own log-in and user account where they can work, download, save and retrieve data in a secure environment. User accounts can be deleted when they become redundant, for example when a user leaves an organization.

Log-on scripts are used to ensure that various drivers, printers and other resources required by a user can all be connected to in a controlled manner. In addition, log-on scripts allow multiple users on a single machine to have access to and connect to the same resources.

Virus scans are critical in ensuring that networks are shielded and protected from internal and external threats. An example of network antivirus protection is that offered by McAfee (Case Study 4.2)

CHAPTER 4

Case Study 4.2

McAfee antivirus protection

McAfee®

Anti-virus

Capture and eliminate threats across your network

Shield your extended enterprise network with comprehensive, multi-tiered virus protection for every system—from desktops, file servers, e-mail servers, and Internet gateways to remote and wireless devices.

File servers and desktops

Comprehensive protection for your network assets McAfee anti-virus products defend your desktops and file servers with complete proactive protection, identifying and capturing the latest exploits, viruses, worms, and Trojans—such as Sasser, My Doom, and Bagel—before they affect your users, systems, and vital corporate data. ▸More

E-mail server

Block inbound threats with e-mail security E-mail is often the culprit for the distribution of malicious programs across your network. With multi-layered protection, our e-mail security protects your e-mail servers against costly network downtime, reduced productivity, and compromised data. ▸More

Internet gateway

Scan and block threats before they reach your network Comprehensive protection begins when you scan all of your inbound and outbound traffic. Our Internet gateway suites, services, and appliances help you stop malicious attacks at the gateway to your network, before they harm your business-critical systems. ▸More

SharePoint protection

Share data without exposure to malicious attacks Computing platforms that support information sharing can leave your enterprise vulnerable to attack. Our collaborative environment products shield these systems from exploits, viruses, worms, and inappropriate content that can wreak havoc throughout your network. ▸More

Information taken from the McAfee website

http://www.mcafee.com/us/enterprise/products/anti_virus/index.html

File cleanups

It is essential that disk space and usage are monitored to ensure that the user does not run out of space, which could be critical for an organization.

Performance monitoring and fault finding

Monitoring the performance of a network and elements such as network traffic and the utilization of resources can ensure a better quality provision because patterns of peak and non-peak usage can be identified, underutilized or overutilized resources can be examined, and solutions put forward to improve the network service.

Fault finding

Tools are available to identify and alert you to faults on the network and provide support in rectifying the faults in some cases.

Activity 4.7

(Tutor-led activity)

1. Discuss and demonstrate how maintenance activities can be conducted.
2. Explain that these sessions have been set up to give students a more practical insight into maintenance activities.
3. Discuss any protocols involved with maintenance activities and ensure that students are aware of the implications of not performing maintenance tasks correctly or sufficiently, in terms of security breaches and lapses in system quality.
4. Some students will be in a position where they may not have engaged with a networked system at a maintenance level previously.
   Engage students in this process by setting up the resources to demonstrate how certain maintenance tasks can be achieved.
5. Set up an environment and demonstrate how to conduct tasks such as creating and deleting user accounts or performing file cleanups. You may have to bring small groups of students up at a time to ensure that they can see clearly what steps are being followed.
6. Provide students with an opportunity to demonstrate their own ability in these areas. Ensure that students take notes of the stages/steps involved. An observation sheet should be used as evidence that these have been completed.
7. The session should end with a review and consolidation, followed by any questions about the activities undertaken.

Documentation

There is a range of documentation that can support network managers and administrators in carrying out network management activities. These can include logging documents such as work, fault or resource logs, or the use of testing documentation such as test plans to ensure that procedures are followed in terms of identifying, recording and resolving issues that may occur across a network.

Configuration options

Configuring a network is an important element to ensure compatibility and consistency across all resources. Configuration options include choosing server and setting rights for user accounts, drive mappings and options for virus scanning procedures.

Questions and review

1. Describe a range of network management functions.
2. Compare and contrast the following networking operating systems:
   - Novell Netware
   - Microsoft
   - Linux.
3. In terms of networking protocols, what is meant by SMTP and SNMP?
4. In terms of network design, why would the following need to be taken into consideration:
   - speed
   - usability
   - functionality
   - cost
   - flexibility?
5. Describe at least three types of network device.
6. What is the function of an 'interconnection device'?
7. Networking tools provide a range of functions and services. Identify two of these.
8. What is meant by 'emerging technologies'? Provide two examples of these.
9. What is the impact of emerging technologies?
10. What maintenance activities should be undertaken on a regular basis?
11. What documentation could be used to support network management activities?
12. Identify a range of configuration options. (One example is 'drive mappings'.)

Assessment activities

Grading criteria	Content	Suggested activity
Pass		
P1	Describe the functions of network management.	Produce a report that incorporates a number of network, fault and performance management tools, techniques and technologies. The first part of the report could include descriptions of the functions of network management.
P2	Describe, with examples, what considerations need to be taken into account when designing a network.	Include a section in the report that describes, with examples, what considerations need to be taken into account when designing a network.
P3	Describe and explain the purpose of a networking tool.	Within the report explain the purpose of a network tool.
P4	Describe and give examples of emerging networking management technologies.	Carry out research into emerging technologies and produce a presentation.
P5	Interrogate a network to identify the network assets and their configuration.	Interrogate a network to identify the network assets and their configuration. An observation or witness statement could be used as auditable evidence.
P6	Undertake routine network management tasks.	In conjunction with P5, demonstrate that you are capable of undertaking routine network management tasks. A checklist can be used that identifies a range of network management tasks.
Merit		
M1	Explain using examples the goals of fault management.	In conjunction with P1, providing examples, identify the goals of fault management within the report.
M2	Describe the potential impact on network systems of emerging networking technologies.	In conjunction with P4, produce additional presentation slides that describe the potential impact on network systems of emerging networking technologies.
M3	Explain how tracking performance variables can be used to modify and improve performance.	Produce an article for submission into an on-line computing magazine that features 'tracking performance variables'. Identify how these can be used to modify and improve performance.
Distinction		
D1	Justify the inclusion of routine performance management activities within a network manager's role.	You have been asked to present information to a group of network technicians and managers who are new to this type of role. Prepare a presentation, electronic or flip-chart based materials that can be visually represented on a VLE. The materials should justify the inclusion of routine performance management activities within a network manager's role.
D2	Explain the need for a proactive network manger to be proactive in their role, giving examples of how such an attitude might be of value.	In conjunction with D1, you could also include information about the need for a proactive network manager and being proactive in their role.

Courtesy of iStockphoto, prill, Image#6201668

An understanding of the IT environment, hardware, software, networks, users and communications is essential if relevant advice and guidance is to be offered to end users.

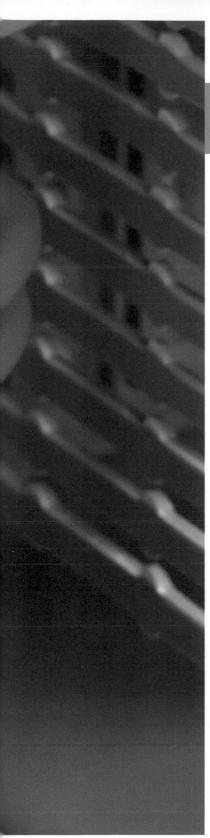

Chapter 5

IT Technical Support

Supporting and maintaining IT systems is a crucial component in any IT implementation strategy. The acquisition of hardware and software and ensuring that systems are compatible, robust and cost-effective needs to be offset against the provisions required to ensure that, once implemented, the system is also supported and maintained either internally or externally.

This chapter will help you to develop an understanding of IT technical support in terms of the skills required to work on a helpdesk. In addition, you should develop an understanding of a range of methods used to support and present information to meet customers' needs.

The chapter is structured around the following learning outcomes:

- Be able to gather information in order to provide advice and guidance.
- Be able to communicate advice and guidance in appropriate formats.
- Understand how the organizational environment influences technical support.
- Understand technologies and tools used in technical support.

Be able to gather information in order to provide advice and guidance

Information gathering is a crucial exercise that is required before any diagnosis, fault finding or monitoring can take place. There are various ways in which information can be gathered and these will be explored in more detail within the next sections.

Information gathering

Information can be obtained through a range of methods, including direct questioning, the use of fault logs, and through diagnostics and monitoring tools. Information gathering can also be conducted via 'syslog' files or the equivalent, which list system messages such as system startup, shutdown and users logging in as administrators.

Direct questioning can be based on a face-to-face approach where you are engaging with a user on a one-to-one or group basis, asking specific questions about their working environment or IT needs. Direct questioning can also take place through the use of a structured questionnaire (Figure 5.1) or an interview.

IT services questionnaire		
Name:		Computer use per day
Department:	Job role:	1 hour or under ☐ 2 hrs ☐ 3 hrs ☐ 4 hours ☐ 5 hours + ☐
What have your experiences been with IT services? Please identify both positive and negative where applicable.		
How often have you used the services of the IT helpdesk in the last month? None ☐ Once ☐ Twice ☐ 3 – 5 times ☐ 6 times + ☐		
What type of query/problem do you usually have? (please tick the appropriate box) Applications ☐ Back-up ☐ Hardware ☐ Network issues ☐ Security issues ☐		
What are the most important qualities that you expect from the IT team? (please rate from 1 – 5) 1 = lowest 5 = highest Friendly team Knowledgeable staff – can answer all queries Quick response time Accessible – different contact methods Flexible – 24/7 support		

Figure 5.1 Example of a structured questionnaire.

Activity 5.1

You work within an IT department for a large public sector organization. You have been asked to create a customer satisfaction survey that will be based on the service and resources offered by your IT department.

1. Identify at least ten questions that would appear on the survey.
2. Why have you selected these questions?
3. Performance levels could be measured in terms of satisfaction of the issue/fault being resolved or the speed and time taken to resolve it. Which would you consider to be of greater importance?

Fault logs provide documentary evidence of errors and faults that occur within an IT environment/system, as shown in Table 5.1. A fault log should contain information such as:

- date of the fault
- details about the fault
- time of the fault
- time taken to fix the fault/resolve the issue
- details about the person logging the fault
- details about the person fixing the fault
- follow-up details and contact information.

Table 5.1 Fault log example

Date of the fault	Fault/problem logged by	Fault/problem description	Fault time	Time taken to resolve	Person who resolved the issue	Follow-up details
30 Apr 08	J. Wilson	System crashed and a document that had not been saved cannot be recovered – has it been lost?	10:45	20 min	R. Owen	Enable the autosave function
30 Apr 08	H. Biggins	Mouse not working	11:03	5 min	R. Owen	Replace with new mouse
30 Apr 08	J. Lemming	Printer toner has run out	11:20	15 min	S. Dean	Replace the toner cartridge

Diagnostics and monitoring software can also be used to detect and gather information about faults and list events. An example of this is 'Event Viewer' (Figure 5.2).

By using a range of information-gathering tools, details can be captured that will allow informed decisions to be made about different types of faults. Diagnosing faults and resolving them quickly are paramount if an efficient and effective IT system is desired within an organization.

When gathering information it is important to identify problems and prioritize them in terms of urgency. High-priority faults may include issues with the network that could impact on hundreds of users, or critical users within an organization. Non-urgent tasks such as

CHAPTER 5

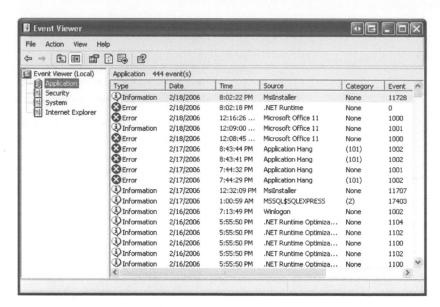

Figure 5.2 Example of an event viewer. http://en.wikipedia.org/wiki/Image:Windows_XP_Event_Viewer.png

restocking paper or changing a toner, in comparison, would be classed as low priority.

Various faults can occur in a system. Some of these can disable certain aspects of a system or network and cause disruption to or loss of service. Some faults may impact on the performance of the system, for example too much network traffic could slow down the system, making it almost inoperable. Other faults may be functional, in terms of a component or piece of software not working or requiring replacement.

Prioritizing faults and identifying their importance is essential. For example, two fault requests may come through, one from a high-end essential user, for example a member of the finance team who has logged a fault with the payroll system that could jeopardize the end-of-month payroll. Another fault could also be logged from a casual user who has requested that the toner cartridge in the printer be changed. Identifying and prioritizing faults should ensure that requests from users who are dependent on the fault being resolved almost immediately, to ensure that they can continue with their job role, are addressed first.

Whatever system faults occur, it is down to the IT support team in terms of first, second or third line support to prioritize the issues and take action to resolve them accordingly.

Validation of information

One important aspect of gathering information is to ensure that it is valid.

Cross-reference checks

The validation of information can be conducted through cross-reference checks with the user to ensure that information collected is a true representation of events. Checks that can be carried out may determine whether:

- a fault occurred, and at what time?
- the account given by the user is a true account of the actual problem
- the fault was an isolated incident or a recurring fault
- the fault only occurs under certain conditions or environments.

By carrying out cross-reference checks the problem may be identified more immediately and isolated a lot more quickly.

Sometimes cross-reference checks are inadequate and the problem will need to be reproduced. The problem may be artificially generated or reproduced on demand, in view of the support person or technician, in order for the fault to be remedied.

FORUMS   Download forum rules

Primed with questions and answers covering six hot categories - from technical queries and general consumer advice to the latest news about the magazine - the PC Advisor forums are here to help. We'll even send you an email when someone responds to your query.

Helproom

Got a technical query or a problem with your PC? With more than a million posts already in our database the chances are good that we've resolved the same problem for someone in the past. Before you post your own thread, try using our search facility - it may save you (and us) some valuable time.

- help burning discs please
- INSTALLING IIS IN XP PRO SP2
- Service Pack 3 (SP3)
- Possible dying hard drive ??
- Home network using mainly D-Link devices
- Corrupt windows
- BIOS failure...
- Registry File Failure

Networking

Want to go wireless and not sure how? Problems with your router? Uncertain about networking hardware? This forum area is strictly for threads about networking - wireless or wired. You ask the questions, we'll try to come up with the answers.

Free Worldwide Wi-Fi access!
Claim your **FREE La FONera Wi-Fi router** worth £30 when you subscribe to PC Advisor for just £19.99 – saving a massive 35% on the newsstand price.

Click here to subscribe now!

Mobile World

Looking for a new mobile phone, satnav or portable internet access device, and want to get the best deal? Perhaps you're happy with your current handset but are mystified by the plethora of features it offers. This forum is for you, as registered users discuss the hot topics, from products and tariffs to 3G coverage and usability.

- Now there's a surprise - Apple learns quickly...
- Cost of Mobile Phone Internet Usage
- Disconnecting 3 Mobile Broadband
- BT Total Broadband Anywhere
- Which basic mobile phone to choose
- N95 and Flip Over
- Google Maps
- n80 feedback
- samsung G600
- Best moblie for reception - advice please

Business

Looking for advice on the best computer hardware and software for your business, and want to get the best deal? PC Advisor's Business forum is the perfect place to discuss the technology issues that affect your company, providing you with a simple way to get the answers you're looking for from like-minded IT professionals.

- Which laptop should I buy?
- Help Document management/scanning
- How to avoid the fraud from internet from agoodic
- Share your marketing techniques.
- Microsoft Office Accounting Express 2008 payroll
- Should we change our router?
- "THEY LIED TV SPOT"
- New Startup Business - Hardware advice
- Fax by Email
- Ufindus.com

Figure 5.3 Forum information.
http://www.pcadvisor.co.uk

Problem reproduction

By cross-referencing information, carrying out checks and monitoring the situation over a period of time it is easier to recognize when and if the problem may be reproduced. Problem reproduction can be monitored on an ongoing basis to assess under what conditions the problem occurs, what triggers the problem and possibly to forecast when the problem may happen again. By setting up a manufactured or controlled environment it may be possible to reproduce the problem so that a solution can be found and implemented.

Websites and forums

As well as being valid, information has to be reliable, and the reliability of information can be measured by reviews from forums (Figure 5.3), where posts can been made by independent members and users. Reliable information can also be obtained from manufacturers' websites (Figure 5.4).

Figure 5.4 Manufacturer information.
http://www.samsung.com/uk/

There can, however, be a vast difference in the quality, reliability and validity of the information depending on the source and their vested interests.

Activity 5.2

1. Compare and contrast two manufacturers' websites and two forums.
2. Select a product or service, for example a printer or new piece of utility software, and use two websites and two forums to gather information and reviews about this item.
3. In your opinion, which site provided the best:

- technical overview in terms of specification
- general product/service review with comments from other users
- product or service general knowledge?

Technical knowledge

Technical knowledge can be acquired from a number of sources (Figure 5.5). Product specifications and manuals provide a great deal of detailed information about a particular product (Figure 5.6).

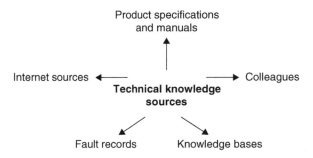

Figure 5.5 Technical knowledge sources.

Product Comparison
Digital Cameras

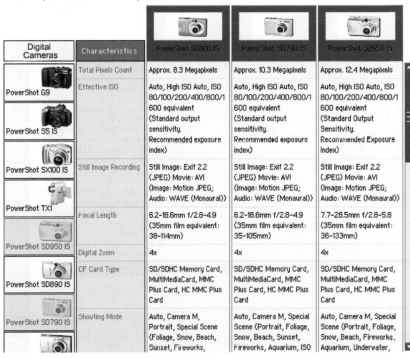

To compare features and specifications, drag the product image to the top of the column. You can select up to three models to compare.

Digital Cameras	Characteristics	PowerShot SD1100 IS	PowerShot SD790 IS	PowerShot SD950 IS
PowerShot G9	Total Pixels Count	Approx. 8.3 Megapixels	Approx. 10.3 Megapixels	Approx. 12.4 Megapixels
PowerShot S5 IS	Effective ISO	Auto, High ISO Auto, ISO 80/100/200/400/800/1600 equivalent (Standard output sensitivity. Recommended exposure index)	Auto, High ISO Auto, ISO 80/100/200/400/800/1600 equivalent (Standard output sensitivity. Recommended exposure index)	Auto, High ISO Auto, ISO 80/100/200/400/800/1600 equivalent (Standard Output Sensitivity. Recommended Exposure Index)
PowerShot SX100 IS	Still Image Recording	Still Image: Exif 2.2 (JPEG) Movie: AVI (Image: Motion JPEG; Audio: WAVE (Monaural))	Still Image: Exif 2.2 (JPEG) Movie: AVI (Image: Motion JPEG; Audio: WAVE (Monaural))	Still Image: Exif 2.2 (JPEG) Movie: AVI (Image: Motion JPEG; Audio: WAVE (Monaural))
PowerShot TX1	Focal Length	6.2-18.6mm f/2.8-4.9 (35mm film equivalent: 38-114mm)	6.2-18.6mm f/2.8-4.9 (35mm film equivalent: 35-105mm)	7.7-28.5mm f/2.8-5.8 (35mm film equivalent: 36-133mm)
PowerShot SD950 IS	Digital Zoom	4x	4x	4x
PowerShot SD890 IS	CF Card Type	SD/SDHC Memory Card, MultiMediaCard, MMC Plus Card, HC MMC Plus Card	SD/SDHC Memory Card, MultiMediaCard, MMC Plus Card, HC MMC Plus Card	SD/SDHC Memory Card, MultiMediaCard, MMC Plus Card, HC MMC Plus Card
PowerShot SD790 IS	Shooting Mode	Auto, Camera M, Portrait, Special Scene (Foliage, Snow, Beach, Sunset, Fireworks,	Auto, Camera M, Special Scene (Portrait, Foliage, Snow, Beach, Sunset, Fireworks, Aquarium, ISO	Auto, Camera M, Special Scene (Portrait, Foliage, Snow, Beach, Fireworks, Aquarium, Underwater,

Figure 5.6 Camera specification comparisons. Canon website: http://www.usa.canon.com/consumer/controller?act=ProductCompareNWAct&fcategoryid=113&modelid5=15652&modelid7=16718&modelid10=16347

CHAPTER 5

Specialist expertise

Some colleagues may have specialist knowledge about a certain product or task requirement. If you work within a team environment there will be a wealth of knowledge and skills that can be shared between colleagues, which can be attributed to an individual's experience or training.

Knowledge bases

Knowledge bases are another source from which technical knowledge can be obtained. An example of a knowledge base is one provided by Microsoft (Figure 5.7), although a range of organizations provide this level of technical help and support.

Microsoft® Knowledge Base Articles on Microsoft FrontPage
Technical Support Information on FrontPage 2000   Bookmark-friendly version

Microsoft support engineers regularly document the solutions to problems that customers call in with. These documents are called Knowledge Base articles. You can search the Knowledge Base database at http://support.microsoft.com/support/search/c.asp. or you can browse the list of topics below.

New Articles

Microsoft Office

- Using FrontPage to Populate a Spreadsheet Component. KB article # 281640.

Permissions

- Applying Unique Permissions to a Subweb. KB article # 205254.

Visual SourceSafe™

- How FrontPage Works With Visual SourceSafe. KB article # 205728.

HOW TO: Use Database Results to Populate the Office Spreadsheet Component in FrontPage 2000

Procedure

1. Open a Web on a Microsoft Internet Information Services Web server.
2. To create the database results page, follow these steps:
   a. Open a new page in FrontPage.
   b. On the **Insert** menu, click **Database**, and then click **Results**.
   c. Select the sample database connection, and then click **Next**.
   d. Under **Record Source**, click **Customers** in the list, and then click **Next**.
   e. Click **Next** to bypass any options.
   f. Select **Table - one record per row**, and then click **Next**.
   g. Select the **Display all records together** button, and then click **Finish**.
   h. Save the page as **customers.asp** to your Web, and then close the page.
3. To create the spreadsheet page, follow these steps:
   a. Open a new page in FrontPage.
   b. On the **Insert** menu, click **Component**, and then click **Office Spreadsheet**.
   c. Save the page as **customers.htm** to your Web.
   d. Click the **Property Toolbox** icon on the spreadsheet control.
   e. Expand the **Import Data** section of the **Property** toolbox.
   f. Type **customers.asp** in the **URL** box.
   g. Click the **Import Now** button.
   h. Select the **Refresh from URL at run time** check box.
   i. Close the **Property** toolbox.
   j. Save the page to your Web again, and then preview the page in Microsoft Internet Explorer.

When users browse the **customers.htm** page via HTTP using Internet Explorer, the Office Spreadsheet Component dynamically loads the data from the database.

Figure 5.7   Microsoft knowledge base. http://www.microsoftfrontpage.com/content/KBarticles/KBarticles.htm. http://support.microsoft.com/kb/q281640/

Fault records can provide a historical account of what has happened previously, from which solutions can be devised based on past experiences. Figure 5.8 provides an example of a completed fault report with the action taken, solution and error codes. In the future a similar fault could be resolved quite easily based on this previous successful diagnosis.

Finally, a range of Internet sources can be used to provide technical information in the forms of frequently asked questions (FAQs) and technical forums.

Microsoft, for example, has 'Technet', a programme and resource that provides technical information, events and news to a range of IT professionals. Microsoft Technet distributes information through their devoted website and specialist magazine, and an open-source blog has also been created.

Microsoft has also established the Microsoft Developer Network (MSDN), which provides a portal of communication directly with developers.

Fault report				
Customer name:	XYZ Company			
Project name:	XYZ System		**Project:**	TNTC10
Project phase:	Development			
Project manager:	Mr. Sample			
Fault name:	Compare doesn't work on directory compare screen		**Fault #:**	20
Test plan/test case #:		2-A	**Severity:**	Severe

Fault report - Source of error codes

Code	Source	Description
FS	Functional specification	Coded per the functional specifications, but the function described in the specifications is incorrect, unclear or incomplete.
DSN	Design	Coded per the design specs, but the design is incorrect, unclear or incomplete. This includes errors in the messages, interfaces and logic.
ARC	Architecture	Selected architectural components do not work together as expected or planned.
HI	Human interface	Coded per specifications, but the human interface (i.e. screens, reports, input documents) defined in the specifications is incorrect, unclear or incomplete.
CD	Code	Not coded according to the detailed design specifications.
PER	Performance	Includes problems with response time, transaction volumes and stress. For example, an on-line query for which 20 second response was required, but an extraordinary test case resulted in a 5 minute turnaround.
DOC	Documentation	The documentation is incorrect (e.g. typos, inconsistencies between the function and the documentation), unclear or incomplete. This includes messages, manuals, on-line help, forms and procedures. The reported fault must be corrected in the appropriate documentation.
ENV	Environment	Test data or current versions of the software and database are not available in the test environment as specified in the standards and procedures manual and in the test plan. The environment must be updated with the required database and software.
DB	Database design	Unexpected results achieved due to the database design (e.g. incorrect primary key definition, incorrect data type). The database must be redefined and the code modified to accommodate the change.

CHAPTER 5

Fault report				
Customer name:	XYZ Company			
Project name:	XYZ System		**Project:**	TNTC10
Project phase:	Development			
Project manager:	Mr. Sample			
Fault name:	Compare doesn't work on directory compare screen		**Fault #:**	20
Test plan/test case #:		2-A	**Severity:**	Severe
Fault description:	**By:**	Ms. Tester	**Date:**	1-1-95

I entered two different directories in directory 1 dialog box and directory 2 dialog box, then selected the compare command button. When I did this the comparison did not take place, i.e. the screen remained as it was.

Assigned to:	Ms. Programmer		
Source of error code(s) (see attached table):	CD		
Action taken:	**By:** Ms. Programmer	**Date:**	1-1-95

The call to the programme that does the comparison (DCComparefiles) was commented out during unit testing and I failed to remove the comments. The comments are now removed.

Solution code(s) (see attached table):	RP	
Configuration item(s) at fault:	DC.FRM	
Retest action	Retest - Ok	

Fault report - solution codes		
Code	**Source**	**Description**
RP	Real problem	Code modified to correct a severe or major error.
CO	Cosmetic	Code modified to correct a cosmetic error, such as spelling or format.
EX	Explanation	Explanation to resolve misunderstanding on the part of the tester.
NP	No problem	Problem could not be repeated and no problem was found.
CR	Change request	The system works according to applicable specifications and a change request is required for system modification.

Figure 5.8 Fault report.Craig Borysowich (IT Toolbox) – fault report posted 30 June 2007, 'Sample Testing Fault Report'. http://blogs.ittoolbox.com/eai/implementation/archives/sample-testing-fault-report-17341.

Activity 5.3

A number of technical knowledge sources is available, such as:

- **product specifications and manuals**
- **human expertise – specialists**
- **knowledge bases**
- **historical fault records**
- **Internet sources – FAQ and technical forums.**

1. **For each of the above knowledge sources, produce a table that identifies the benefits and drawbacks of the source.**

Be able to communicate advice and guidance in appropriate formats

Providing IT technical support and being able to communicate this to end-users effectively is essential if issues are to be resolved easily.

End-users

End-users can fall into a number of categories. Users may be new to a system and require training; these can be classed as 'novice'. Some end-users have built up a wealth of experience and have undergone training; these can be classed as 'experienced'.

Another differentiation that can be used between end-users is to define them in terms of their specialism or technical expertise; these can be described as being 'technical' or 'non-technical' end-users.

When IT support is being given to end-users the type of communication used should reflect their technical ability and level of experience. For example, there is no point in communicating via e-mail to inform a user that a technical problem has been resolved if they are incapable of engaging with e-mail.

Activity 5.4

1. Search through IT job adverts and select suitable job examples for novice, experienced and specialist/technical users.
2. What criteria differentiate the three categories?
3. What, if any, common criteria or qualities are required across all three jobs and categories?

Types of advice

The type of advice given will be dependent on the nature of the issue or problem. The types of advice that can be given could include:

- recommendations for repair or replacement of components
- provision of training or direct instruction
- bug fixes
- installation of patches
- system reset or reboot.

As components become obsolete or fail they may need to be replaced. Advice in terms of hardware or software requirements and specifications may be required from key members within an IT team who would be able to identify the most suitable replacements, costs, services and provisions.

In addition, any procurement policy or service-level agreements that may be in place would need to be addressed as part of the advice given.

Training of end-users or providing direct instruction may be an essential part of any communication provided to support the implementation of new hardware, upgrading of systems, installation of software or any other IT-related activity.

Changes, maintenance, fixes, installations or reboots all need to be communicated to end-users, and possible advice on the impact that these will have on certain activities or short-term tasks needs to be given. If users are informed of any disruptions to their working patterns because of essential repairs or upgrades that will inevitably improve their working environment they will be more receptive and appreciative of the tasks undertaken.

Activity 5.5

You work on the IT helpdesk for a large training company. A number of requests have been received and these have been logged. Unfortunately, your colleague has gone off for lunch and has forgotten to order the requests in terms of priority.

1. Look at the following list of requests and order them in terms of what you consider to be priority, with the most urgent requests at the top of the list.

Request	Priority
Printer has run out of paper	
System has crashed that has affected the payroll run in the finance department, which needs to be completed within the next 4 hours	
A user has mentioned that their mouse is not working and may need cleaning	
There has been a security breach and a virus has been detected on the network	
A server has gone down, which means that the sales department cannot function	
Six new members of staff require new network log-ins	

2. Discuss your priority list with another group/class member to see what you have agreed on and where the differences are.
3. Identify another three possible end-user requests.

Communications

The way that you communicate to end-users in terms of the type of communication tool used and the way that the communication is delivered is very important in terms of offering reassurance and support.

A range of communication methods can be used to update and inform end-users about developments and IT support queries. If there is a physical helpdesk facility, end-users may engage with a support person face to face to discuss problems. E-mail correspondence is also an effective way of logging problems and responses to queries.

Information of a more generic nature can also be communicated as a secondary provision through newsletters or FAQs (see Case Study 5.1) posted on an intranet site or a forum.

Case Study 5.1

Microsoft Support Lifecycle Policy FAQs

1   What is the Support Lifecycle policy?

The Microsoft Support Lifecycle (MSL) policy standardizes Microsoft product support policies for Business and Developer products, and for Consumer, Hardware, and Multimedia products. The Support Lifecycle policy was originally announced on October 15, 2002. A Support Lifecycle policy update went into effect June 1, 2004. The Support Lifecycle policy update applies to most Business and Developer products that were in Mainstream support on June 1, 2004, and to future product versions. The new Support Lifecycle policy provides:

Business and Developer products

Microsoft will offer a minimum of 10 years of support for Business and Developer products. Mainstream Support for Business and Developer products will be provided for 5 years or for 2 years after the successor product (N + 1) is released, whichever is longer. Microsoft will also provide Extended Support for the 5 years following Mainstream support or for 2 years after the second successor product (N + 2) is released, whichever is longer. Finally, most Business and Developer products will receive at least 10 years of online self-help support.

Consumer, Hardware, and Multimedia products

Microsoft will offer Mainstream Support for either a minimum of 5 years from the date of a product's general availability, or for 2 years after the successor product (N + 1) is released, whichever is longer. Extended Support is not offered for Consumer, Hardware, and Multimedia products. Products that release new versions annually, such as Microsoft Money, Microsoft Encarta, Microsoft Picture It!, and Microsoft Streets & Trips, will receive a minimum of 3 years of Mainstream Support from the product's date of availability. Most products will also receive at least 8 years of online self-help support. Microsoft Xbox games are currently not included in the Support Lifecycle policy.

To find the support timelines for your product, visit the Select a Product for Lifecycle Information site (products listed by Product Family) or the Support Lifecycle Index site.

2   Does this policy affect U.S. customers only, or is this policy global?

The Microsoft Support Lifecycle (MSL) policy is a worldwide policy. However, Microsoft understands that local laws, market conditions, and support requirements differ around the world and differ by industry sector. Therefore, Microsoft offers custom support relationships that go beyond the Extended Support phase. These custom support relationships may include assisted support and hotfix support, and may extend beyond 10 years from the date a product becomes generally available. Strategic Microsoft partners may also offer support beyond the Extended Support phase. Customers and partners can contact their account team or their local Microsoft representative for more information.

CHAPTER 5

3   What is the difference between Mainstream Support, Extended Support, and online self-help support?

Support provided	Mainstream Support phase	Extended Support phase
Paid support (per incident, per hour, and others)	X	X
Security update support	X	X
Non-security hotfix support	X	Requires extended hotfix agreement, purchased within 90 days of mainstream support ending.
No-charge incident support	X	
Warranty claims	X	
Design changes and feature requests	X	
Product-specific information that is available by using the online Microsoft Knowledge Base	X	X
Product-specific information that is available by using the Support site at Microsoft Help and Support to find answers to technical questions	X	X

Note: A hotfix is a modification to the commercially available Microsoft product software code to address specific critical problems.

4   Will Microsoft offer support beyond the Extended Support phase?

Microsoft understands that local laws, market conditions, and support requirements differ around the world and differ by industry sector. Therefore, Microsoft offers custom support relationships that go beyond the Extended Support phase. These custom support relationships may include assisted support and hotfix support, and may extend beyond 10 years from the date a product becomes generally available. Strategic Microsoft partners may also offer support beyond the Extended Support phase. Customers and partners can contact their account team or their local Microsoft representative for more information.

http://support.microsoft.com/gp/lifepolicy

When information is being communicated to end-users consideration should be given to how it is delivered, rather than just the communication tool being used.

Some users may be quite distraught when they log a problem with IT support: it could be that they have lost an important document or the system has crashed in the middle of an urgent task. As a result of this they will expect somebody at the end of a telephone or face to face to be quite supportive and sympathetic to their needs. Being empathetic and reassuring the end-user may calm them down and ensure that the problem is resolved effectively and quickly.

Activity 5.6

A number of soft skills can be used to reassure or calm down a user.

1. If you were working on a helpdesk as first line support how would you deal with a very irate user who had lost all of the work they had input in the last two hours?

2. Can you identify five essentials skills and qualities that a helpdesk person should have?
3. Why do you think that these skills are important for this job role?

Checking solutions

Once a solution has been communicated and the problem has been resolved, there may be a need to follow this up with further checks and reviews to ensure that it does not happen again.

Testing may be required as a reactive response to see whether the fault has been resolved and that the system is now performing as it should. If the initial problem was software based, upgrades or further configurations may be required. If it was hardware based, further testing may be required to ensure compatibility, reliability, and so on. Tests can also be used proactively to identify any further issues within a system.

Feedback from users is important to ensure that the advice given was sufficient and correct, and that the solution was successful or successfully implemented. One way of doing this is to ask users to complete a review document or customer satisfaction form.

Understand how the organizational environment influences technical support

Technical support can be influenced by a number of factors, one of which is the environment in which an organization operates.

Working procedures and policies

Different organizations have different policies and procedures for dealing with operational issues such as reporting faults, security, service-level agreements, confidentiality and sensitivity of information, and other areas of IT support.

Organizational guidelines are usually set out in the form of a policy, an agreement or some other communicable statement, possibly posted electronically for all users to access. Within these guidelines there may be information about how to use certain resources, for example e-mail (Case Study 5.1). In addition, there may be instructions about how to report a fault or use of the Internet.

Case Study 5.2

Example of e-mail guidelines

Email Usage Guidelines    connect™

Revision Date: 09 March 2005

Aim

The key to good email usage is care and thought. It is important to realise that the actions of each user can have a dramatic effect on the speed and reliability of the system for the whole company.

The main effect of careless usage is on data storage and bandwidth usage. Your company has a finite level of each, and it is surprisingly easy to use it all up. It is possible to expand the capacity, but this could cost many thousands of pounds and should only be done if necessary.

This guide is designed to provide you with some simple steps to ensure that you are using email responsibly. It is aimed at users of Outlook, but much of the advice also relates to other email systems.

Guide

1   Be organised

Outlook allows you to maintain your own structure for organising data. This can cut down on wasted space, and time finding important emails. You can create as many folders as you need under your inbox.

To add a folder, simply right-click on your **Inbox** and select **New folder**.

To move an email into this new folder drag it with the left mouse button.

Sample folders   *Adding a new folder*

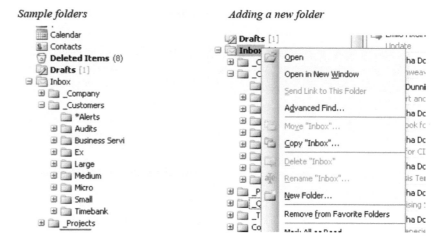

2   Sent items

Sent items also account for a large amount of data and can easily be forgotten if you do not refer to them regularly. The same method can be used for filing sent items, either as a subfolder of the sent items, or in the folders you created above.

http://www.connect.co.uk/pdfs/Emailusageguidelines.pdf

Service-level agreements

Service-level agreements outline the responsibilities and commitments with suppliers or contractors stipulating the service that they are offering and to what level of standard/quality. Service-level agreements can have an impact on technical support for a number of reasons.

First, the agreement may dictate the boundaries and accountability for any repairs or activities carried out on hardware or software within an organization.

Secondly, the technical support may, under the conditions of the agreement, be outsourced to a third party provider.

Finally, any repairs or activities that take place may have to conform to certain rules or conditions. Non-compliance could result in invalidation of the terms and conditions of any agreement made.

Confidentiality and sensitivity of information are usually communicated within a range of policy documents that may also include references to legislation such as the Data Protection Act and Computer Misuse Act.

Outsourcing and geosourcing

The culture of an organization can change depending on the provision of services, whether these are outsourced or geosourced, or whether services such as IT support are provided in-house. Outsourcing can be beneficial to an organization under the right conditions (see Case Study 5.2). Geosourcing is based on finding the best skills in different geographical locations, possibly driven by costs, favourable exchange rates, lower resource costs or the ability to source specialist skills and expertise.

Case Study 5.3

Example of outsourcing

Outsourcing

Here's how outsourcing increased our competitiveness

Anthony Alan Foods Ltd is a leading supplier of low-fat cakes and pastries, sold under the Weight Watchers brand. Established in Barnsley in 2001, it now employs 30 people and supplies over 40 different product lines to major supermarkets. Outsourcing has always been integral to the company's strategy, as operations director Matt Carr explains.

Matt Carr
→ **Anthony Alan Foods Ltd**

What I did

Know your strengths
"The company was set up to capitalise on consumer demand for appetising, lower-calorie equivalents of popular bakery products. We knew from the outset that we would target supermarket chains. That meant we needed sizeable production capacity, which we didn't have.

"We also wanted the ability to introduce new products quickly and to diversify into other food sectors. We knew that our expertise lay in product development, food technology and marketing, rather than production. We also recognised that we would need to devote considerable resources to managing customer relationships, since supermarkets expect a high level of involvement from suppliers."

Matt's top tips:
- "Communicate openly at every stage."
- "Don't expect things to be perfect from the start."
- "Learn from your suppliers."

Set expectations up front
"In the early months, we wasted a lot of time with suppliers through mismatched expectations. Nowadays, we always provide a 'ways of working' document up front that clearly sets out our goals and expectations."

Take care with outsourcing overseas
"Some of our outsourcing partners are based overseas, which has worked well. However, we were a bit naïve to start with and should have been more aware of the cultural differences involved. More research would have helped."

CHAPTER 5

Website: Business Link: http://www.businesslink.gov.uk/bdotg/action/detail?r.
l1=1074404796&r.l3=1073921035&type=CASE%20STUDIES&itemId=1076888409&r.
l2=1074456652&r.s=sc

Activity 5.7

1. Based on the information provided in Case Study 5.2, what do you consider to be the positives and negatives of outsourcing?
2. Are there any services that you feel should not be outsourced in an organization, and if so why?

Organizational constraints

A number of organizational constraints can impact on technical support, such as the cost of resources, time and user expertise.

Depending on the size of the organization, the cost of resources can have a tremendous impact on technical support. The larger the organization the more resources are required to support its infrastructure, and the more hardware, software, communication tools and people are needed to manage and support the system.

Within a large organization there could also be issues with time as more users place greater demands on support people for troubleshooting and repairs.

The level of support provided can depend on the people providing it, i.e. user expertise. If an IT department has relatively new and/or inexperienced people, this may reflect on the level of support that they can offer to end-users, especially as the environment and systems may be unfamiliar to them.

Understand technologies and tools used in technical support

A range of technologies and tools can be used to aid IT support personnel in their job roles. Some of these include the use of e-mail or diagnostic tools. In addition, looking to the future and analysing various trends, for example, moves towards remote support or outsourcing may enhance the level of service provided by IT support.

Technologies

The use of e-mail to communicate to users and to receive technical information regarding faults and problem solving can be effective and timely. An electronic fault log can be retained and updated at various stages throughout a repair, and feedback can be provided to the user on progress made.

E-mail distribution lists can be set up for groups of users to facilitate communication of certain issues, for example a distribution list for all users within a functional area or with a specific role or responsibility. Distribution lists are more focused and direct than company-wide e-mails.

Software diagnostic tools such as WinVNC can be used for technical support tasks. VNC (virtual networking computer) is remotely controlled software that allows a user to view one computer and interact with another over the Internet using a VNC 'viewer' and a VNC 'server'. WinVNC is a useful freeware remote management tool. If a PC has the VNC server component running on it, the small client programme can be used to connect to the PC via remote access/dialup.

Another example of a remote access system is 'logmein.com', a provision that offers a range of remote support functions and facilities.

Remote diagnostic connections will allow connection and diagnosis to take place offsite and away from the system where the fault diagnosis is taking place.

The control panel also offers a range of technical support functions (Figure 5.9). These include the time and date, printer, mouse and keyboard configurations, and password properties. User accounts can also be set up to protect and restrict log-ins to certain applications.

GUI desktop and display set-up   Printer, mouse and keyboard
                                              configurations

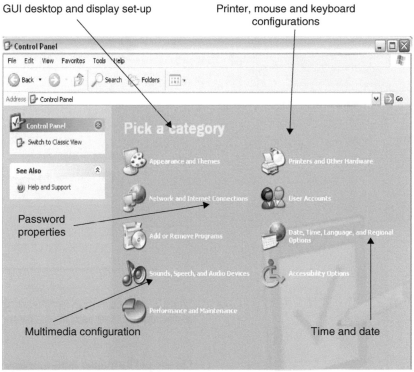

Password properties

Multimedia configuration   Time and date

(a)

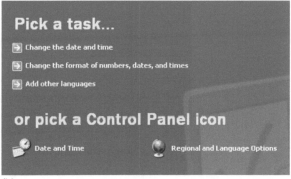

(b)

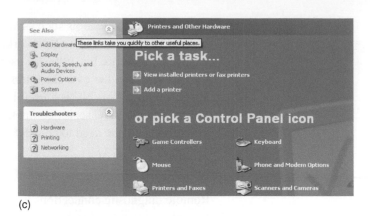

(c)

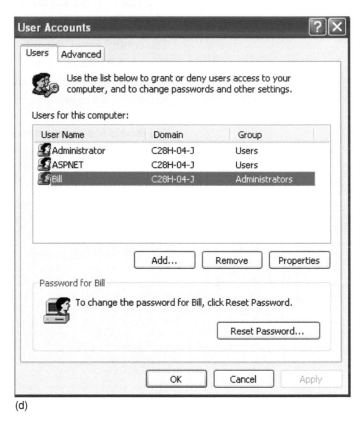

(d)

Figure 5.9 (a) Control panel features and functions; (b) time and date settings; (c) printer configurations; (d) user accounts and password properties.

On the performance and maintenance menu in the control panel (Figure 5.10) you can also carry out a range of scheduled tasks such as backing up data.

Future trends

Over recent years many organizations have moved part or all of their IT service and technical support provision to third party remote providers. The benefits of remote support can range from reduced costs and overheads to receiving a more specialist service because of a wider skill set and range of technical expertise. Problems can be monitored more accurately and constantly and errors may be diagnosed and eliminated more rapidly.

CHAPTER 5

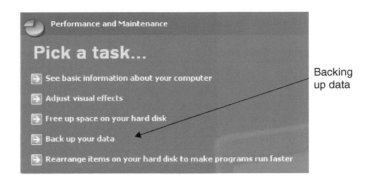

Backing up data

Figure 5.10 Performance and maintenance menu.

Trends have also emerged in terms of the movement towards developmental systems that analyse and report on faults for other users, such as those planning corporate training programmes.

Activity 5.8

1. Carry out research into an organization that has or is using a remote support provision.
2. Describe what benefits the organization experienced as a result of this provision.
3. Identify any limitations of the remote support that the organization experienced.

Questions and review

1. Identify a range of information-gathering techniques.
2. Why is it important to validate information? Identify two ways in which you can evidence validation of information.
3. What are the advantages and disadvantages of using an 'open user forum' to gather and validate information?
4. Provide two examples of a knowledge base and provide another example of how technical knowledge can be obtained.
5. Differentiate between the following end-users:
   * novice
   * experienced
   * technical.
6. Communicating advice and guidance in appropriate formats is extremely important in terms of providing an IT technical support service. Provide three examples of types of advice that can be given.
7. Communications can be initiated in a number of ways when providing advice and guidance. Provide two benefits of communicating in the following formats:
   * direct with a user – face to face
   * using e-mail
   * contributing to a FAQ or forum site.
8. In what ways can solutions be checked?
9. In what ways can the organizational environment influence technical support?
10. In terms of organizational constraints, how could user expertise impact on or influence technical support?

CHAPTER 5

Assessment activities

Grading criteria	Content	Suggested activity
Pass		
P1	Identify different sources of technical information for problems and for each, describe how valid they might be.	Produce a table that identifies different sources of technical information for problems. Describe how valid they might be.
P2	Appropriately respond to questions raised by users and check that solutions proposed were successful.	Ask your tutor or a friend for a checklist of appropriate questions, then write down appropriate solutions in response to them. The responses could be checked by another class member to see if the proposed solutions were successful.
P3	Explain information gathering techniques that can be used to answer a problem or request.	Produce a short presentation covering information gathering techniques that can be used to answer a problem or request.
P4	Describe different communication routes through which advice and guidance can be made available to users.	In conjunction with P3, additional slides could be produced that describe different communication routes through which advice and guidance can be made available to users.
P5	Describe two current diagnostic and one current monitoring software tool used by support staff and outline possible future developments in this area.	Produce a short report that addresses a range of technical issues, such as diagnostic tools and monitoring software. The report could describe two current diagnostic tools and one current monitoring software tool used by support staff and outline possible future developments in this area.
P6	Describe how organisational policies and procedures impact on the provision of technical advice and guidance.	Produce an information leaflet aimed at providing advice to middle management within a technical support environment. The leaflet should include information about organisational policies (P6), outsourcing (M3) and advances in support systems (D2). The first part of the information leaflet could describe how organisational policies and procedures impact on the provision of technical advice and guidance.
Merit		
M1	Evaluate different sources of technical information.	In conjunction with P1, produce a written evaluation about the different sources of technical information identified within the table.
M2	Produce appropriate support material that will guide users in relation to a specific area of expertise.	Produce a user guide that will support users in relation to a specific area of expertise.
M3	Explain the advantages and disadvantages to users and organizations of outsourcing the provision of technical support.	In conjunction with M2, devote a section within the leaflet to explain the advantages and disadvantages to users and organizations of outsourcing the provision of technical support.
Distinction		
D1	Evaluate the impact of an organisation's policy in relation to the support service provided to internal customers.	In conjunction with P5 an evaluative section could be included within the report that looks at the impact of an organisation's policy in relation to the support service provided to internal customers.
D2	Review comprehensively a recent advance in support systems technology and evaluate the impact it is having on the provision of such support.	The final part of the leaflet should demonstrate that you have comprehensively reviewed a recent advance in support systems technology. Evaluate the impact it is having on the provision of such support.

Courtesy of iStockphoto, Nikada, Image# 6220160

As IT systems become more advanced the level of support required to troubleshoot and repair these systems also increases. Troubleshooting and repair can save an organisation time and money if faults are detected and resolved both quickly and efficiently.

IT Systems Troubleshooting and Repair

One of the many roles of an IT department and helpdesk support is that of IT troubleshooting and repair. The ability to diagnose faults quickly and accurately is a skill that is valued and requires a range of technical knowledge and expertise.

This chapter will support you in developing a range of skills for diagnosing and troubleshooting hardware and software problems. A range of health and safety considerations will also be provided to ensure that practical activities are undertaken in a safe working environment.

The format of this chapter will be structured more around the practical application of skills and knowledge. It will be very much activity based, to provide a foundation from which you can demonstrate any theory that has been acquired throughout this book and other information sources.

The chapter is structured around the following learning outcomes:

- Know how to identify and select suitable remedies to repair IT systems.
- Be able to apply fault remedies to hardware and software systems.
- Understand how organizational policies impact on diagnosis and repair.
- Be able to apply good working practices when working on IT systems.

Know how to identify and select suitable remedies to repair IT systems

There are a number of ways and remedies that can be used to repair IT systems. Some of these ways are based on information provided by third parties, while others are more practically focused and examine the fault history or previous symptoms. Once the fault has been diagnosed the procedure to remedy it can be administered.

Identifying and selecting remedies

Identify and select remedies by using:

- knowledge bases
- technical manuals
- Internet, e.g. FAQs and discussion forums, manufacturers' websites
- colleagues
- training programmes undertaken
- fault history.

Types of remedies

A number of remedies can be used and applied in the process of repairing IT systems. They will differ depending on whether or not the fault is hardware or software related. These remedies may include:

- repairing or replacing the hardware
- fxing communication paths
- reconfiguring the software
- applying a software patch
- reinstalling software
- instructing the user in terms of correct equipment usage.

Repairing or replacing hardware

Repairing or replacing hardware may be the only option if other avenues of fault detection have been followed. Repairing components may be possible if they can be accessed easily and if there are parts within them that can be removed, cleaned and replaced.

With the falling costs of some components, sometimes it is easier to replace an item and also more cost-effective, especially taking into consideration factors such as labour resources and downtime of the system.

Fixing communication paths

Communication paths and links between systems are paramount, for example linking to the Internet, sending e-mails and sending data across a network. Some communication paths are facilitated through the use of cables, therefore a physical cabling fault may be the cause of the fault. Other communication path errors could be due to software settings or obstacles that can block signals, such as wireless interference.

Software patches

Software remedies can range from reconfiguration, installing a patch or a complete reinstall, depending on the nature of the problem.

Software patches are small pieces of software/programmes that are used to fix problems within a piece of software. Patches can be used to fix bugs and improve the performance, efficiency and usability of the original software. Patches can also be used to ensure that systems security is not compromised, as illustrated in Case Study 6.1.

Case Study 6.1

Ubuntu patches flaw

Ubuntu became the latest Linux vendor to patch a vulnerability in the open-source operating system's kernel that could have left the door open for hackers to find their way into users' machines.

In an email sent last night, the Linux vendor warned users to update all machines running recent versions of Ubuntu, ranging from 6.06, which was released back in mid-2006, to version 8.04, which came out earlier this year. The problem also applied to other versions of Ubuntu such as Kubuntu, Edubuntu and Xubuntu.

Ubuntu administrators wrote in the email: 'It was discovered that there were multiple NULL-pointed function de-references in the Linux kernel terminal handling code. A local attacker could exploit this to execute arbitrary code as root, or crash the system, leading to a denial of service.'

The email also detailed a number of other bugs which could be exploited by an attacker who already had some level of access to a computer running Ubuntu.

A number of other Linux vendors including Novell have recently released similar patches to address the problems.

Original article: Ubuntu issues security patch for kernel flaw from ZDNet Australia

By Renai LeMay

Published: 26 August 2008 09: 09 GMT

http://software.silicon.com/os/0, 39024651, 39275144, 00.htm
http://www.zdnet.com.au/

CHAPTER 6

Activity 6.1

1. Why was the patch released? What was its purpose?
2. Can you find two other case study examples of patch releases and identify what the patches were used for?

Reinstalling software

Uninstalling a previous version of a piece of software and then reinstalling it is another way to overcome and remedy software issues such as crashes or errors that occur within the programme.

The reinstallation process will involve the user in carrying out a series of stepped procedures, as shown in Case Study 6.2 with a snapshot of reinstalling Windows® XP.

Case Study 6.2

Start of reinstallation procedures for Windows XP

1. Insert the Operating System CD.
2. Shut down the computer, and then turn on the computer.
3. Press any key when the Press any key to boot from CD message appears on the screen.
4. When the *Windows XP Setup* screen appears, press <Enter> to select *To set up Windows now.*
5. Read the information in the *License Agreement* window, and then press <F8> on your keyboard to agree with the license information.

Instructing the user

Instructing the end-user either through training or by directing them to manuals and/or step-by-step guides is another way of overcoming issues. If a user does not understand a particular system task, providing them with support and guidance may be a more effective way of remedying the problem.

Activity 6.2

A number of remedies can be applied in the repair of IT systems.

1. Demonstrate that you can reconfigure a piece of software.
2. Demonstrate that you know the procedures for uninstalling and then reinstalling a piece of software.
3. Produce a one-page user guide that identifies at least two faults and how they can be remedied (one should be a hardware fault).

Nature of reported faults

Faults can range in complexity. Some faults are easily remedied and have a transparent and identifiable solution. Other faults, however, can be more complex with non-specific symptoms, making them more difficult to identify and remedy.

Attenuation – the reduction of signal strength over a given distance.

Near-end cross-talk (NEXT) – interference between two pairs in a cable measured at the same end of the cable as the transmitter (http://en.wikipedia.org/wiki/Near_end_crosstalk).

Time-domain reflectometry (TDR) – a measurement technique used to determine the characteristics of electrical/transmission lines.

Activity 6.3

For each of the following scenarios, identify a range of typical faults that could be the cause. For each fault identified, include a proposed remedy.

1. You have sent information to a printer but nothing is printing out.
2. You created a document a few days ago using applications software but now, when you go to open it, you cannot find it.
3. You are trying to access a site on the Internet but you cannot get a connection.
4. A piece of software that you installed last week keeps crashing.
5. Data that you are trying to send over the network is not being received.

Be able to apply fault remedies to hardware and software systems

Various tools, techniques and troubleshooting methods can be used to remedy a range of hardware and software faults.

Hardware tools and techniques

Tools that can be used to support an end-user or a technician in the repair of a hardware component range from testing instruments, monitoring devices and carrying out self-test routines to physical tools such as screwdrivers, pliers and a torch.

One electrical/electronic test instrument that is commonly used in conjunction with IT system diagnosis and repair is that of portable appliance testing (PAT) (Figure 6.1). When a piece of electrical equipment has passed a PAT test, a 'pass' sticker (Figure 6.2) is applied to the component or item to verify that it has passed.

Digital multimeters can be used to measure voltage, current, resistance, capacitance and cable continuity. Multimeters check the physical connectivity.

Cable testers can be used to check physical connectivity. The types of cables that can be checked include: foil twisted pair (FTP), unshielded twisted pair (UTP), coaxial, twinaxial and 10-BASE-T. Testing equipment can also be supplied for fibreoptic cable.

A cable tester can be used to:

- test and report on the conditions of the cable in terms of noise, attenuation, near-end cross-talk (NEXT) and far-end cross-talk (FEXT)
- carry out functions such as traffic monitoring, wire mapping and time-domain reflectometry (TDR) tasks
- provide information and statistics about network traffic, utilization and packet error rates, as well as provide limited protocol testing.

Self-test routines

When you switch on a computer, the first programme, which consists of a set of instructions that are kept in the computer's read-only memory (ROM)

CHAPTER 6

Figure 6.1 PAT tester. https://www.pat-services.co.uk/martindale-hpat-600-handypat-932.htm

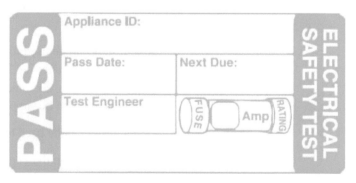

Figure 6.2 PAT pass label. http://www.pat-test-labels.co.uk/pat_testing_pass_labels.html

will run. The systems hardware is checked to ensure that everything is functioning accordingly. This preboot sequence, which occurs in computers, printers and routers, is known as a power-on self-test (POST).

Monitoring devices

A monitoring device is another tool that can be used to track, assess and report on faults that can be identified in hardware and software systems.

Monitoring devices can be used proactively to alert a user or an IT department about an anomaly within the system, or can be used as a reactive measure to track occurrences such as network usage, Internet access or data exchanges between users.

Suitable tools

When embarking on a hardware repair or troubleshooting activity, physical tools such as screwdrivers, pliers and torches are required to remove and replace components.

Software tools and techniques

A range of software tools and techniques that can be used to identify, monitor and remedy systems faults. These include the use of diagnostics software such as antivirus, test utilities, error-logging and monitoring programmes, and also a range of system-specific applications.

Diagnostics software

Diagnostics software can be used to optimize the efficiency or performance of a system by detecting areas of weakness or inefficiency. It can also be used to detect errors or defects within a system, and to support users and specialists/experts in their job role to make more informed decisions, as illustrated by Case Study 6.3.

Case Study 6.3

Medical Diagnostic and Treatment Software Holds Potential to Save Lives and Improve Patient Care Worldwide

http://www.microsoft.com/en/gb/default.aspx

Sagestone Consulting and the Robertson Research Institute chose the Microsoft .NET Framework and Visual Studio .NET as the foundation for NxOpinion, a medical decision-support software programme that delivers up-to-date information about hundreds of diseases and ailments to physicians.

REDMOND, Wash., Jan. 21, 2004 – Nearly 100,000 people die in U.S. hospitals each year due to diagnostic errors, according to the Agency for Healthcare Research and Quality. After his friend's 21-year-old son died in an emergency room as a result of a misdiagnosis, Michigan neuropharmacologist Dr. Joel C. Robertson resolved to create a technology product that might prevent similar tragedies by helping doctors to more quickly and accurately assess a patient's condition.

'In emergency medicine, you may have only a few minutes to diagnose a patient and start a treatment,' says Robertson. 'When I reviewed the medical records on my friend's son, I uncovered some additional evidence that likely would have saved his life if it had been available to the doctors who treated him – but it took me two weeks to dig out the answer.

'That's when I decided that someone needed to develop a diagnostic decision-support software program that could be made available globally, at little or no cost

CHAPTER 6

to any physician who needs it, and that could be constantly updated by experts from around the world.'

Robertson channeled his vision into the creation of NxOpinion (Nx is short for 'Next'), a real-time diagnostic tool built on Microsoft technology – and with assistance from Microsoft Research – that promises to provide physicians with timely and relevant information concerning hundreds of non-chronic illnesses and diseases. The non-profit Robertson Research Institute (RRI), which Robertson founded in February 2001, plans to distribute NxOpinion at no cost to doctors and medical organizations in areas where this type of resource is most needed – such as impoverished and developing regions, and rural clinics. An initial version of the software and associated database of medical information is slated to launch in 2005.

RRI selected Sagestone Consulting Inc., a Microsoft Gold Certified Partner, to develop NxOpinion. In turn, Sagestone chose to build the software on a foundation of Microsoft technologies that includes the .NET Framework, Microsoft Visual Studio .NET 2003, Microsoft Windows Server 2003 and Microsoft SQL Server. After two months of setting the vision for NxOpinion, Sagestone began constructing it in January 2003.

The Microsoft technology behind this initiative was vital in terms of its ability to extend the different NxOpinion applications across different platforms that doctors might use in the field, as well as for its reliability,' says Keith Brophy, chief executive officer of Sagestone. 'It allowed us to create a sophisticated and powerful diagnostic product, yet still make it easy for even the most non-technical physician to use. This project is a wonderful affirmation of what you can accomplish with Microsoft's platform and tools.'

Designed for use on a Tablet PC or desktop computer, NxOpinion prompts a physician to enter details about a patient's condition as if he or she were describing the case to a colleague. Based upon each new piece of information, NxOpinion suggests possible ailments and asks for other evidence – for example, 'Have you taken a blood pressure reading in the leg?' – until the doctor is confident that he or she has pinpointed the most likely diagnosis and identified a viable treatment option.

Physicians can adapt NxOpinion to their skill level, language and level of available medical resources. For example, if the physician does not have the optimal equipment or expertise to run a lab test or perform a textbook-perfect medical procedure, NxOpinion can suggest viable alternatives. Since the .NET Framework and SQL Server store all their information in Unicode, which is a way of consistently representing character sets across the world, Sagestone has been able to build multi-language capability into NxOpinion without the added development time and expense of converting data from one character set to another for each language.

'Two of the most important design goals for NxOpinion were to make it easy for physicians to use and allow them to find any piece of information in three mouse clicks or less,' says Brophy. 'Microsoft Windows Forms technology allowed us to develop a very rich and interactive interface that delivers information to the user on one screen. With Windows Forms, we're also able to cache a very large and diverse body of medical information right on the physician's device for instant retrieval.'

The core technology that will enable NxOpinion to intelligently weigh different pieces of evidence and generate the next logical question or suggestion is a Bayesian inference engine currently being constructed by the Sagestone and Microsoft team using the .NET Framework and C# programming language. Dr. David Heckerman of Microsoft Research, who is one of the foremost experts in Bayesian decision theory, contributed to the overall design of the diagnostic engine.

Named for the 18th-century English mathematician Thomas Bayes, Bayesian logic essentially deals with calculating the statistical probability of different outcomes based upon prior observations. The NxOpinion diagnostic engine is programmed to correlate the different pieces of available medical evidence, compare them to the disease profiles stored in the NxOpinion knowledge base and use the weightings assigned to each piece of information to generate a differential diagnosis – a list of potential disease candidates.

'The physicians whom we most want to reach with NxOpinion initially are the ones working in small, isolated communities and villages,' says Robertson. 'Often, they're the only doctor for miles around, and they don't have access to laboratory equipment or specialists in all these different areas of medicine. Because of its unique ability to mirror the way these physicians think, adapt to their real-life working conditions and prompt them with new ideas to consider, NxOpinion has the potential to save thousands of lives a year.'

RRI has raised more than $9 million in philanthropic contributions to cover initial development of the NxOpinion end-user application, the NxOpinion Content Creator and the Bayesian diagnostic engine. RRI plans to enlist the help of other charitable organizations in distributing the software in disadvantaged areas, and the institute also hopes that a few major technology equipment manufacturers will step forward to provide free or discounted computer hardware for physicians who need it to run NxOpinion.

'When I think about the return on investment for NxOpinion,' says Robertson, 'it's measured not in dollars but in the lives we're going to help save.'

Sample screenshots to illustrate how NxOpinion helps physicians rapidly formulate medical diagnoses and treatment options.

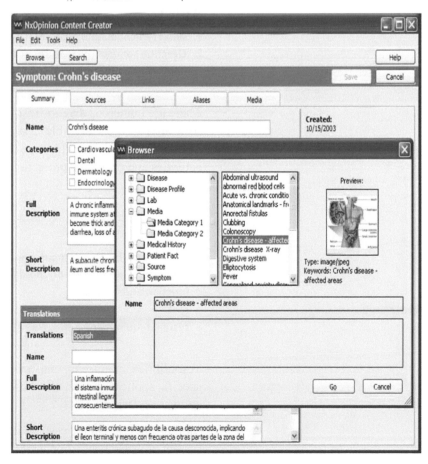

CHAPTER 6

Microsoft.com

http://www.microsoft.com/presspass/features/2004/Jan04/01-21NxOpinion.mspx

Activity 6.4

1. What type of diagnostic system has been developed?
2. Describe some of the features of the system and how it can be used by doctors and other medical specialists.

There are numerous diagnostic, utilities, monitoring and error-logging software packages available to support users and technicians in identifying and fixing problems, checking performance, monitoring activities and providing protection against viruses and spam. One of these is shown in Figure 6.3.

Activity 6.5

1. Produce a table that compares and contrasts four other types of diagnostic packages. The comparison should take into consideration:
- function and features
- cost
- system requirements.

Test utilities, monitoring and error-logging programmes and system-specific applications can all be used to remedy faults in hardware and software systems.

Monitoring programmes can be used to track and record either the system or users' behaviour. Areas that can be monitored include

Figure 6.3 Example of diagnostics software. http://www.dl4all.
com/category/diagnostic_software/

web monitoring, key logging, screen snapshot and e-mail capture.
Monitoring software can all be used to monitor the frequency and traffic
usage on networks and automatically log any anomalies or errors.

Using troubleshooting techniques

Various troubleshooting techniques can be applied to hardware and
software systems. Troubleshooting consists of a series of events or
stages, each one trying to eliminate or resolve the issue.

Substitution

One method of troubleshooting is 'substitution', where components
are removed and exchanged to identify where the fault is. By using a
systematic approach, errors can be eliminated.

Testing

Testing hardware and software can ensure that faults are detected before
a system goes 'live'. Testing can also ensure that any faults in the future
can be detected early on before they become more serious.

If a fault is detected isolated testing can eliminate other components
within the system. If a methodical approach is used testing can be
carried out until the fault is discovered and rectified.

Changing

Changes can be made to both the hardware and software of a system.
Substitution can be used to replace like for like, whereas change could
require a completely different component solution. Examples could
include changing brands/manufacturer, e.g. from an nVidia graphics
card to an ATi branded one, or an interface change from integrated drive

electronics (IDE) to serial advanced technology attachment (SATA), or simply changing an internal modem for an external one.

Activity 6.6

1. Identify four other changes that can be made to either hardware or software within an IT system.

Reinstalling software

Sometimes software can fail or crash. One way to overcome this is to reinstall the software. A reinstallation can:

- f x corrupt files by replacing the files with the original ones
- return the settings to default, for example changing back a language setting
- replace missing files, if they have been accidentally moved or deleted
- replace shared library files.

Elimination

A process of elimination can identify and isolate components that may have failed or be suspected as the source of a fault. By removing individual components and carrying out a series of checks the elimination process can be used as an effective and efficient means of fault diagnosis.

Applying bug fixes

If a bug arises in a piece of software or operating system, vendors can be contacted or websites can be checked to obtain a 'bug fix'. A bug fix is essentially a rewritten piece of updated code that should remedy any bugs or faults identified.

Generating error codes

Error codes are generated by software failure conditions. Case Study 6.4 identifies a range of common error messages.

Case Study 6.4

Common error messages

Common error messages

Windows installation and startup errors

Error: Non-system disk or disk error. Replace and press any key when ready.

Remedy: This happens when the computer can't read the information on a floppy disk or the PC's hard disk. You may have left a floppy disk in the drive by accident; removing it and then pressing any key should solve the problem. If not, there may be a problem with the hard disk. Try pressing any key, restarting the computer, or switching it off for a few minutes before starting up again. If none of these remedies work, you may have a serious hard disk problem and you'll need to contact the computer or hard disk manufacturer, the shop where you bought the PC or hard disk, or a qualified technician.

Error: Fatal error: An error has been encountered that prevents setup from continuing. One of the components that Windows needs to continue setup could not be installed.

Remedy: This usually means that Windows XP could not read data on the setup disc. Try taking the disc out of the drive, cleaning it (by wiping a dry, clean, soft cloth in a straight line from the inside to the outside of the disc, not round in a circle), and trying again. If this doesn't work, you may have to get a replacement disc.

Error: System has recovered from a serious error.

Remedy: If you see this message every time you start the computer, it means an error is stuck in the computer's memory. Correct it by clicking on Start and right-clicking on My Computer, then choosing Properties. Click on the Advanced tab, then on the first Settings button under Performance. Click on the Advanced tab, and click on Change. In the next dialogue box, select No Paging File, then click on Set. If you get a warning message, click on Yes or OK. Then click on 'System managed size' and on OK. Close all the dialogue boxes and restart the computer.

Anthony Dhanendran, Computeractive 14 Oct 2004
http://www.computeractive.co.uk/computeractive/features/2014063/common-error-messages-stamp-problems-part

---

Understand how organizational policies impact on diagnosis and repair

Organizational policies can impact on IT systems hardware and software and repair. Policies may be set up internally that dictate the conditions in which a repair is to take place, when, where and by whom. An organization may be tied into certain service-level agreements (SLAs) that can limit the amount of diagnostics and repairs that can be carried out on components.

An organization will also have to comply with a range of legislation that may also influence the way in which diagnosis and repair can be carried out.

Customer issues

Communications

Good communication channels are essential before, during and after any diagnosis or repair of IT systems. The diagnosis or repair of any hardware or software can be dependent on the testimony of an end-user or a third party such as a customer. Their interpretation of a fault can influence the diagnosis procedure and also postdiagnose any course of action such as a repair or replacement.

Understanding the impact of diagnosis and repair

The impact of the diagnosis or repair should also be taken into consideration in terms of how this will affect the end-user or customer, for example. If a fault has been reported, the symptoms may indicate that the fault is non-complex and can easily be remedied within a matter of minutes or hours. This type of diagnosis and repair would cause little disruption to the end-user in terms of continuing with their tasks or job role. If, however, the fault was intermittent or complex, then a series of tests and diagnostic tools and techniques may have to be used to decide the cause and appropriate remedy. These types of faults could have a detrimental impact on the end-user, especially if their system keeps randomly crashing or stops being effective. In these situations a replacement component or system may be substituted in the interim whilst tests are carried out to rectify the fault, as and when it occurs.

Customer handover and acceptance process

Once the repair has been completed a process of customer handover and acceptance should follow that details the nature of the fault and the remedy, for example changing of component or reinstalling a piece of software. The physical handover procedure may also be accompanied by a follow-up feedback form or an e-mail that acts as closure to the fault.

Unresolved faults can be quite frustrating for end-users and they can potentially impact on the service to external clients and customers. If a user is reliant on a fully operational system and a fault occurs this can delay certain tasks. If a job role is wholly dependent on the use of a computer to access and process data, even the smallest of faults could cause major problems and delays to services.

External considerations

External considerations that can impact on IT systems diagnosis and repair include:

- legislation and legal issues
- level agreements
- escalation procedures
- documentation and reporting.

Legislation and legal issues

Legislation and compliance with legislation can have a profound effect on organizations and policy making. Legislation such as the Health and Safety Act and the Data Protection Act may dictate the conditions in which a repair can be carried out. It can also stipulate the safety equipment required to undertake a repair and the status of a technician or qualified IT employee who is undertaking the repair.

If the repair results in the recovery, storage or transfer of data, legislation may also dictate how, by whom and where this is carried out. Employees are also bound by legally binding contracts, so issues of confidentiality and professionalism throughout the repair process need to be addressed and upheld.

Level agreements

Level agreements or SLAs outline the conditions, expectations and obligations in terms of the level of support offered. SLAs may include provisions for the type of service, timings and response times, resource arrangements, security, provisions and contingencies for system downtimes.

Case Study 6.5 provides information about Datanet, an SLA provider.

Case Study 6.5

Datanet.co.uk – service-level agreement provider

Service Level Agreement

Home > About Datanet > Service Level Agreement

About Datanet

Vacancies
Terms & Conditions
▸ Service Level Agreement
Infrastructure
Downloads
Datacentre Diary
Awards and Affiliates

A Service Level Agreement (SLA) is an informal contract between a provider and a customer that explains the terms of the providers responsibility to the customer and the type and extent of remuneration if those responsibilities are not met.

Datanet's Service Level Agreements are explained in a simple, clear language and are designed to have a tight focus on your business. Our Service Level Agreements are available below in PDF format.

- Co-located & Dedicated Server SLA (142 KB pdf)
- Rack Space SLA (142 KB pdf)
- IP Transit SLA (143 KB pdf)
- Private Circuit SLA (139 KB pdf)
- DSL SLA (107 KB pdf)
- Premier Support SLA (103 KB pdf)

Datanet are also a member of CISAS, an Ofcom approved dispute resolution service. See our Terms and Conditions for details.

Contact Us

T: 0845 130 6010
F: 0845 130 6020
E: Solutions@datanet.co.uk

Downloads

- Co-location and Dedicated Server SLA
- Rack Space SLA
- IP Transit SLA
- Private Circuit SLA
- DSL SLA
- Premier Support SLA

Premier support
Service level agreement

Datanet service level agreement

1. General

　1.1 This document is a service level agreement (SLA) setting out the levels of services to be provided by Datanet to the Customer under this agreement and compensation for failure to meet those service levels.

1.2 In this SLA a reference to a paragraph, unless stated otherwise, is a reference to a paragraph of this SLA.

1.3 In this SLA words, abbreviations and expressions have the meanings given in the Datanet Master Service Agreement. General Terms and Conditions except as set out below:

Availability	All the time in any calendar month for which the network and any service equipment is not subject to any service affecting faults, and is therefore Available. Business Day shall mean every day excluding Saturdays and Sundays and national holidays in England
CDR	Means the committed data rate for each port set out in the service Order Form and provided as part of a Service.
Fault	Shall mean a material defect, fault or impairment in a service, which causes an interruption in the provision of the service
Non-Service Affecting	Means not materially affecting the performance or quality of the service
Service Affecting	Means causing full or partial loss of the ability to transmit data
Third Party System	Means a telecommunication system that is neither owned nor operated on behalf of Datanet

1.4 Datanet reserves the right to amend the SLA from time to time. The latest version of the SLA will always be posted on the Datanet website.

2. Hardware Guarantee

2.1 Datanet guarantees the functioning of all hardware components covered under Premier Support and will replace any failed component at no cost to the customer. Hardware replacement will begin once Datanet identifies the cause of the problem is hardware related. Datanet will replace the faulty equipment next business day with a similar preconfigured device.

2.2 The customer may arrange a courier to collect the replacement unit in order to expedite delivery, costs involved in doing so being the sole responsibility of the customer.

3. Remote Management

3.1 Our engineers will maintain and troubleshoot the Firewall as necessary via secure VPN. We will manage and support remote user configurations and branch to branch VPNs as part of our managed firewall solution. This will cover activities such as:

Adding, removing and configuration of IPSEC VPNs
Port forwarding of services to internal devices
Policy additions/modifications and deletion
Reporting

4. Software Guarantee

4.1 Your firewall will be remotely upgraded and reconfigured, keeping your security up to date and extending the functionality of the hardware. Upgrades will be performed under the following conditions:

CHAPTER 6

A software update is announced which enhances or resolves security issues within the device.
Additional functionality is available which will benefit the customer or offer increased performance.

5. Fault handling/Response Time Agreements

5.1 Datanet offers response time agreements, during the business day (8.30am to 5:30pm, Monday to Friday) as follows:

You can call us on our 'lo-call' number 0845 130 6010 and expect a prompt answer, you will be able to speak to a member of the technical team who will be familiar with your account and services. You can expect to be able to speak to a network engineer normally straight away and always within 2 hours.

5.2 Replacement hardware will be shipped to the customer providing the fault can be identified and a replacement configured before the 17:00 postal deadline.

http://www.datanet.co.uk/sla.aspx

Escalation procedures

Escalation procedures provide a framework of stepped procedures that can be initiated if service levels do not meet a required standard. This framework can include determining faults and identifying problems in the repair or reporting stages. It could also be initiated by problems not being resolved within a certain time-frame.

Documentation and reporting

Documentation is crucial in any diagnosis, fault-reporting or repair. An account of faults should be logged to assist in the diagnosis of any fault. A record of events can also determine whether the fault is isolated to a given component or programme, or is widespread throughout a system or network. The log or audit trail of previous faults can also determine the complexity of the issue and the resolution method.

Activity 6.7

Fault diagnosis and repair can be isolated to a single user or component; however, the impact of the diagnosis and/or repair could have wider implications at an organizational level.

1. **In groups, discuss four different types of faults, hardware and software, and identify whether or not the extent, diagnosis and repair of these faults could impact on the user and the organization.**
2. **Within your group, carry out research to identify what the minimum requirements are within an SLA. Provide another example of an organizational SLA.**

Organizational considerations

When any diagnosis or repair is being undertaken, although the problem may be quite contained to a single user or component, a range of wider organizational issues may need to be considered.

Security

Security issues can impact on the initial fault, diagnosis and repair process.

A range of security procedures and policies that identify protocols in terms of good working practice and that outline the consequences of system misuse can prevent some faults and problems from occurring. The spreading of viruses, Trojans or worms throughout a system, or infiltration by unauthorized users, can be addressed through a range of firewall, spyware and antivirus software.

Having an IT policy that is available to all employees, outlining good working practice and the use of IT systems, can also prevent some system faults from occurring through misuse or ignorance.

Costs

Several factored and unfactored costs are associated with the diagnosis and repair of IT systems.

Many larger organizations have an internal IT provision that will maintain, monitor, update and provide ongoing support to users. Other organizations use third party organizations to provide remote support and backup, which may be offered at a more competitive rate than having an internal IT infrastructure.

Other factored costs include the replacement of components and the purchasing of newer software versions. These should be included within any procurement procedures.

Unfactored costs may result in system downtimes and services not being met for third parties such as customers if a fault is ongoing or severe enough to disable the organization for a period of time, as highlighted in Case Study 6.6. In these situations costs may need to be allocated for more resources in terms of expertise, external support or the acquisition of new systems, as opposed to spending valuable time rectifying and troubleshooting the existing ones.

Case Study 6.6

Sainsbury's online problems

The Risk Factor
Software failures and successes dissected daily

Sainsbury's On-Line Problems Cost It At Least $2M

J Sainsbury plc

The UK's third largest grocer Sainsbury's has finally fixed a software problem that brought down its on-line grocery service, according to Vnunet.com. The problem which kept Sainsbury's from being able to process customer orders, started the evening of the 17th of June. Customers who had placed orders discovered that they were canceled, which left many rather unhappy.

As reported by the Independent,' Sainsbury's delivers to 90,000 customers a week and their average spend is thought to be around £80 per transaction … the technical glitch has already cost the grocer well over £1 m in lost online sales.'

Sainsbury's competitors were quick to jump on its problems, with ASDA, for example, offering 'free delivery on home shopping when [customers] enter a promotional code.'

The embarrassment factor of the problem was also magnified as a result of some unfortunate timing. Sainsbury's chief executive Justin King was boasting to City analysts about the growth of its online site during a first quarter results presentation on the morning of the 18th of June, just as the problem was becoming public.

'The online operation is continuing to perform well, with sales growth at over 40 per cent,' King reportedly said.

I am sure King was rather incensed not to be informed about the problem before he gave his presentation.

Sainsbury's has offered the approximately 20,000 shoppers affected by the problem a £10 voucher as compensation.

ieee spectrum online, 20 June 2008
http://blogs.spectrum.ieee.org/riskfactor/2008/06/sainsburys_online_problems_cos.html

Activity 6.8

1. What was the cause of Sainsbury's online problem?
2. How did this impact on Sainsbury's, its competitors and customers?

Impact of system downtime

The impact of system downtime can be catastrophic for some organizations, as illustrated by the Sainsbury's case study with a cost of £1 m (~$2 m).

At a user level, downtime could prevent people from carrying out certain tasks as part of their job role. At a functional department level, certain procedures may be delayed that could result in the loss of customers and potential new business. At an organizational level, system downtime could have critical effects. Millions of pounds can be lost, customer loyalty could be compromised, and the overall reputation and corporate image of the company could be put into question in terms of quality and reliability of provision.

Contractual requirements

Employees are bound by the terms and conditions of their contracts. In relation to the diagnostics and repair of IT systems, the environment in which these repairs are carried out should fall within the boundaries of the contract. This could be in reference to working hours, adhering to certain policies and legislation or abiding by certain codes of conduct, or acting professionally if working offsite on a customer's premises.

CHAPTER 6

Trend analysis of faults reported

By maintaining fault logs and incident report sheets, analysis can be carried out to see when, where and how certain faults have occurred, the diagnosis procedure and outcome. Trend analysis may be able to support technicians and IT personnel who carry out repairs in predicting faults in certain areas and also in conducting future repairs more efficiently and specifically.

Resource allocation

Resources include physical equipment such as hardware and software. Within an IT environment they also include the people who provide support and who carry out repairs and systems maintenance.

Within an organization, utilizing resources and allocating them to areas that require ongoing support, possibly to critical areas such as network management, may need to be considered and enforced. Ensuring that operational tasks on a day-to-day basis are being carried out is essential to support stakeholders and other functions and services both internal and external to the organization.

Prioritization of jobs

As system faults are reported a need arises for the prioritization of support and repairs. Certain jobs can be classed as essential or critical, for example a server going down and preventing users from accessing the network. Other faults such as a printer not functioning properly can be classed as less critical and therefore may be prioritized a lot lower.

Jobs can also be prioritized in terms of essential users, therefore certain users may be considered to be high priority, so any faults reported by them would be addressed first and lower priority users would be retained and logged in date and time order.

Be able to apply good working practices when working on IT systems

The communication, application and enforcement of good practice are essential when working with and on IT systems. Good practice can cover a range of areas such as health and safety, having awareness about the conditions and environment in which a repair is being carried out, and adhering to general working policies and procedures.

Health and safety

When you are using equipment to carry out fault diagnosis or repair you should first ensure that the equipment you are using is appropriate and used correctly.

Correct use of equipment

The type of equipment that may be used includes screwdrivers, pliers, test meters, electrical equipment, and antistatic protection – wrist-straps or mats.

Electrostatic discharge

The most effective way of avoiding electrostatic discharge (ESD) is to use a wrist-strap or grounding mat. Other measures and precautions that can be taken to avoid ESD include:

- Remove any jewellery.
- Try to avoid wearing any clothing that may conduct ESD, such as woollen items.
- Remove any cords away from the back of the computer.
- Avoid working with computers in extreme weather conditions such as during an electrical storm.
- Continuously touch unpainted metal surfaces to ensure that you and the computer are at zero potential.
- Try standing rather than sitting, especially because the chair can generate more electrostatic charge.

Manual handling procedures

Good working practices should also apply to manual handling procedures. Manual handling broadly includes the transporting or supporting of a 'load'. This can include moving, lifting, pushing or using any hand or body force to achieve this. In terms of IT systems this may involve the moving and lifting of computer systems, or the setting up of an IT room to accommodate these systems.

Considering fire safety

Fire safety should be prioritized as a major health and safety issue. Set procedures with regard to taking appropriate steps in the event of a fire should be communicated to all users.

Exit areas should be clearly identified and access paths kept free of any obstructions. Boxes and equipment should not be stacked or stored in these areas. In an event of a fire, users should also be aware of who to contact and, if appropriate, the necessary equipment to use, if trained as a fire warden/marshal.

Correct disposal of old parts and equipment

There are certain procedures to follow when disposing of old parts and equipment. Many organizations have policies and procedures in place regarding the recycling or redistribution of components.

Laws that are in place regarding the safe disposal of systems can also dictate how disposal will be implemented, and organizations that do not adhere may be prosecuted and fined.

CHAPTER 6

Activity 6.9

1. Carry out research to identify two organizations that have embraced a recycling policy for the safe disposal of old parts and equipment.

Considering health and safety of other people

Within an organization there are potential health and safety hazards that could occur and impact on other people within the same working environment. Consideration in terms of using laser equipment, keeping work areas clutter free, and ensuring that leads and wires are safely tied or trailed in appropriate conduits, etc., should be a matter of priority to avoid trip hazards or other electrical dangers.

Availability of first aid and supervision

Within any environment there should always be appointed and known first-aiders and supervision to provide medical assistance to users and employees who may have been involved in a health and safety incident.

Working practices

Good working practices should be adhered to when working on IT systems.

Obtaining permissions before repairing

Permissions should be obtained from end-users, customers and owners of systems who have identified the need for a system repair.

Permissions may be given verbally if a repair is to take place internally within a department for an employee. However, more formal permissions may be required if repairs are carried out on a third party's system, for example a customer.

A disclaimer may need to be signed if the system holds confidential or sensitive data. In addition, the owner/user may have to sign to agree to the repair and the conditions of the repair in terms of estimated time, or approve the necessary course of action once the fault has been detected.

Preparing the worksite

The worksite or environment in which the repair is taking place may have to be prepared to comply with health and safety regulations and requirements. In addition, certain objects such as mobile phones may not be allowed into server rooms because of the electromagnetic radiation.

Recording information

Information regarding the diagnosis and repair process should be recorded to identify elements such as product keys to activate the software, licence numbers, installation date, etc. This information can be used to create a profile of a certain component or system that may be required in the future.

Data backup

Under certain conditions (if feasible) data should be backed up before any repair. Damage to the hard disk or errors made in terms of reformatting the drive or a software installation could corrupt or delete data files.

In addition, if the system is removed offsite, a risk assessment should identify the chances of the system being stolen (portable device), damaged or lost.

Maintaining security and confidentiality of data

Data security and confidentiality of data can be preserved through a number of physical measures, such as setting up passwords, using backup procedures, and having antivirus and spyware or monitoring software installed.

When working with IT systems, procedures such as setting up screen passwords or locking the keyboard when you are away from the system and in an environment where other users could gain access would also be considered good working practice.

The working environment of users of ICT is of major importance when it comes to looking at issues such as health and safety. In addition to the considerations identified, a number of other factors should be considered that cover environmental, social and practical aspects of working conditions (Figure 6.4).

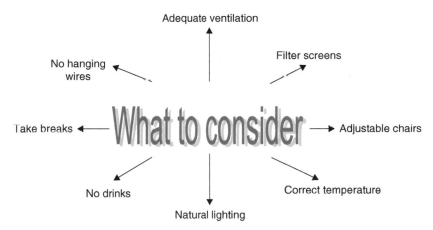

Figure 6.4 Health and safety considerations.

Users should be working in an environment that has adequate ventilation and natural lighting, and the temperature should be conducive to a computing environment, especially as computers give out large quantities of heat.

Computer users should also have sufficient support peripherals such as filter screens to minimize glare and height-adjustable chairs. When working at a computer no food or drink should be consumed in case liquid or crumbs fall onto the keyboard or into the case. Wires should always be packed away in appropriate conduits and not left trailing across the floor.

The best measure for health and safety in the workplace is to use common sense and adhere to standard ways of working. Organizations also offer guidelines and procedures for maintaining good working practice.

CHAPTER 6

Activity 6.10

You have been employed in a new role as a health and safety administrator within a large insurance company. You work with employees from a range of departments including call centre representatives, IT specialists and human resources.

1. Design a health and safety leaflet that could be used to introduce and support new employees in their job roles.
2. Design a short presentation that focuses on a range of good working practices that can be applied when working on IT systems.

Questions and review

1. What tools and techniques are available to support technicians and support people in identifying and selecting remedies for IT systems?
2. What are the benefits of using a manufacturer's website for fault diagnosis and system remedy solutions?
3. Select and identify three types of remedy for either a hardware or a software fault.
4. Provide two examples of a complex fault that has non-specific symptoms.
5. Identify two electrical/electronic test instruments that can be used to remedy a system fault.
6. What are the main features of diagnostic software?
7. Identify four types of troubleshooting technique.
8. Provide four specific examples of how organizational policies can impact on diagnosis and repair.
9. In what situation might you be required to enforce 'escalation procedures'?
10. Why should an organization need to consider the impact of system downtime as part of the diagnosis and repair process?
11. What health and safety considerations need to be factored in when conducting a diagnosis or repair to an IT system?
12. What would you consider to be good working practice when working on IT systems?

Assessment activities

Grading criteria	Content	Suggested activity
Pass		
P1	Troubleshooting complex IT problems and identify suitable remedies.	Think of a complex IT problem and demonstrate that you can troubleshoot it by identifying possible problems and outlining solutions for them. This practical activity can also be linked into P3 and P4.
P2	Describe the use of two hardware and two software tools to troubleshoot complex IT problems.	Produce a report that describes the use of two hardware and two software tools to troubleshoot complex IT problems.
P3	Apply a fault remedy safely to a complex hardware problem.	As part of the practical activity set up to meet the conditions of P1, you could also apply a fault remedy to a complex hardware problem.
P4	Apply a fault remedy to a complex software problem.	As part of the practical activity set up to meet the conditions of P1, you could also apply a fault remedy to a complex software problem.
P5	Describe how organisational policies impact on the troubleshooting and repair process.	In conjunction with P2 you could include a section within the report that describes how organisational policies impact upon the troubleshooting and repair process.
P6	Communicate effectively with users during fault diagnosis activities.	Demonstrate that you have communicated effectively throughout the fault diagnosis activities by orally explaining to the observer what tasks you are undertaking throughout. You should also be capable of answering any questions that might be raised by the observer during this process.
P7	Apply good working practices when troubleshooting an IT system.	In conjunction with P1 you should demonstrate that you have applied good working practices during the troubleshooting process.
Merit		
M1	Define an appropriate remedy for an identified hardware fault from a range of possible solutions giving reasons for their choice.	In conjunction with P3 you could also define an appropriate remedy for an identified hardware fault from a range of possible solutions, giving reasons for your choice.
M2	Define an appropriate remedy for an identified software fault, from a range of possible solutions giving reasons for their choice.	In conjunction with P3 you could also define an appropriate remedy for an identified software fault from a range of possible solutions, giving reasons for your choice.
M3	Analyse the appropriateness of a range of hardware and software tools when troubleshooting an IT problem.	In conjunction with P2 and P5 you could include a section within the report that analyses the appropriateness of a range of hardware and software tools when troubleshooting an IT problem.
M4	Explain and justify good working practices when troubleshooting and repairing an IT system.	In conjunction with P7 you could also provide a written explanation that justifies good working practices when troubleshooting and repairing an IT system.
Distinction		
D1	Detail the potential impact of two types of faults on the specific user of the system, on the organisation and any external customers.	In conjunction with M1 and M2 you could also detail the potential impact of two types of faults on the specific user of the system, on the organisation and on any external customers.
D2	Explain with examples how appropriate organisational guidelines and procedures can help to minimise the impact of IT faults.	Make an information sheet that provides examples and an explanation as to how appropriate organisational guidelines and procedures can help to minimise the impact of IT faults.

CHAPTER 6

Index

T - #0942 - 101024 - C164 - 280/208/8 - PB - 9780750686532 - Gloss Lamination